# RAPID PRODUCT DEVELOPMENT

# RAPID PRODUCT DEVELOPMENT

**M ADITHAN**
Professor Emeritus
Department of Mechanical Engineering
VIT University, Vellore
Tamil Nadu

**NEW AGE INTERNATIONAL (P) LIMITED, PUBLISHERS**
New Delhi • Bangalore • Chennai • Cochin • Guwahati
Hyderabad • Kolkata • Lucknow • Mumbai
Visit us at www.newagepublishers.com

Published by New Age International (P) Ltd., Publishers
**First Edition:** 2015



**BRANCHES**

- **Bangalore** 37/10, 8th Cross (Near Hanuman Temple), Azad Nagar, Chamarajpet, Bangalore-560 018
Tel.: (080) 26756823, Telefax: 26756820, **E-mail: bangalore@newagepublishers.com**
- **Chennai** 26, Damodaran Street, T. Nagar, Chennai-600 017
Tel.: (044) 24353401, Telefax: 24351463, **E-mail: chennai@newagepublishers.com**
- **Cochin** CC-39/1016, Carrier Station Road, Ernakulam South, Cochin-682 016
Tel.: (0484) 2377303, Telefax: 4051303, **E-mail: cochin@newagepublishers.com**
- **Guwahati** Hemsen Complex, Mohd. Shah Road, Paltan Bazar, Near Starline Hotel, Guwahati-781 008
Tel.: (0361) 2513881, Telefax: 2543669, **E-mail: guwahati@newagepublishers.com**
- **Hyderabad** 105, 1st Floor, Madhiray Kaveri Tower, 3-2-19, Azam Jahi Road, Near Kumar Theater Nimboliadda, Kachiguda, Hyderabad-500 027, Tel.: (040) 24652456, Telefax: 24652457
**E-mail: hyderabad@newagepublishers.com**
- **Kolkata** RDB Chambers (Formerly Lotus Cinema) 106A, 1st Floor, S.N. Banerjee Road, Kolkata-700 014
Tel.: (033) 22273773, Telefax: 22275247, **E-mail: kolkata@newagepublishers.com**
- **Lucknow** 16-A, Jopling Road, Lucknow-226 001
Tel.: (0522) 2209578, 4045297, Telefax: 2204098, **E-mail:lucknow@newagepublishers.com**
- **Mumbai** 142C, Victor House, Ground Floor, N.M. Joshi Marg, Lower Parel, Mumbai-400 013
Tel.: (022) 24927869, Telefax: 24915415, **E-mail: mumbai@newagepublishers.com**
- **New Delhi** 22, Golden House, Daryaganj, New Delhi-110 002
Tel.: (011) 23262368, 23262370, Telefax: 43551305, **E-mail: sales@newagepublishers.com**

**ISBN: 978-81-224-3810-9**

**₹ 499.00**

**C-14-10-7987**

Printed in India at Glorious Printers, Delhi.
Typeset at Aakriti Graphics, Delhi.

**PUBLISHING GLOBALLY**
**NEW AGE INTERNATIONAL (P) LIMITED, PUBLISHERS**
7/30 A, Daryaganj, New Delhi-110002
Visit us at **www.newagepublishers.com**

# Preface

Development of a product involves carrying out a network of related activities. These may have cycle times which may be constant or widely varying and some precedence relations may exist among them. Due to this complex nature of the product development cycle, accurate insight into the happenings at any time is difficult. Hence it takes a long time to launch products. However, need to compress product development time drastically is vital for the survival of the organizations and for their success. Because, time-to-market greatly enhances the competitive advantages of any organization.

Rapid Product Development (RPD) is a philosophy used for shortening product development time. RPD is the synergic integration of the various time compression technologies — which may be virtual or physical in nature. Time compression can be achieved by doing the various design, testing and manufacturing activities in the virtual world as far as possible. When this becomes no longer possible, one can shift to the physical world where all out efforts are made to do them as fast as possible. It is of course important to ensure that the activities happen in synergy so that the benefits achieved at any of them are carried forward till the end.

Higher Educational Institutions (HEIs) and R & D centres have been in the fore front and traditionally have been catalysts in the absorption of new technologies in the industries.

When new technologies emerge, be it CAD, CAM, CNC or Rapid prototyping (RP), institutions and R & D centres took the initiative to diffuse these technologies into the industries. The scope of RP and Rapid Tooling has enlarged into Rapid Product Development and more and more industries will be tooled up with these technologies to meet the challenges posed by the global economy.

This book is an attempt to bring in the awareness of RPD to our designers and engineers. We hope the product designers and engineers working in R & D centres and industries will be immensely benefitted by this book.

**M. Adithan**

# Contents

# Part I

## INTRODUCTION

1. Product Development Cycle
2. Influence of Innovations on Product Development
3. Rapid Product Development — An Overview

# CHAPTER 1

# Product Development Cycle

## 1.1 PRODUCT DEVELOPMENT CYCLE

With the outbreak of industrial revolution the trend of large scale production for domestic and foreign markets has set in. Pioneer countries in this trend envisaged for better products through huge investments in R&D and could manage to be always ahead of others. The competition has been intensified by globalization and trade liberalization. To sustain present competition, companies world over are compelled to develop new and better products. It is also important that the manufacture is economical and the product reaches the market ahead of the competitors.

The huge investments required for new product development is associated with a very high risk as a product's success or failure in the market depends on its acceptance by the user. For any progressive company it becomes essential to take this risk to retain the market position and to survive in the present day competition, which makes it even more important to ensure commercial success of the product. New products have very little chance of survival without proper *Unique Selling Propositions (USPs)* attached to it. With growing awareness, consumer's choices are more governed by the price-features offered with the product.

The various **definitions of design** are given below:

- Finding the right physical components of a physical structure
- Engineering design is the use of specific principles, technical information and imagination in the definition of a mechanical structure, machine or system to perform specified functions with the maximum efficiency
- A creative activity that involves a process of bringing into being something new and useful that has not existed previously.

Most of these classical descriptions refer to the ingredients of Design Process, but not of designing (Figure 1.1). We prefer to think of design as a process that initiates change in the man made things which leads to the new experiences most appropriate for the situation, brings out the real meaning to the designers and at the same time offers benefits to the users. When such experiences become part of culture, they create a foundation for the significant real world change. The act of designing, as of now is more concerned not with the end product, but with the changes that manufacturers, distributors, users and the society as a whole have to make, in order to adapt and to get benefit from the new design (Figure 1.2).

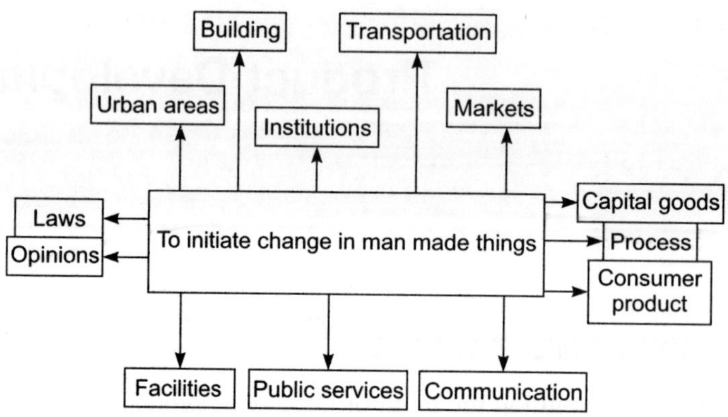


**Figure 1.1** Factors influencing product design

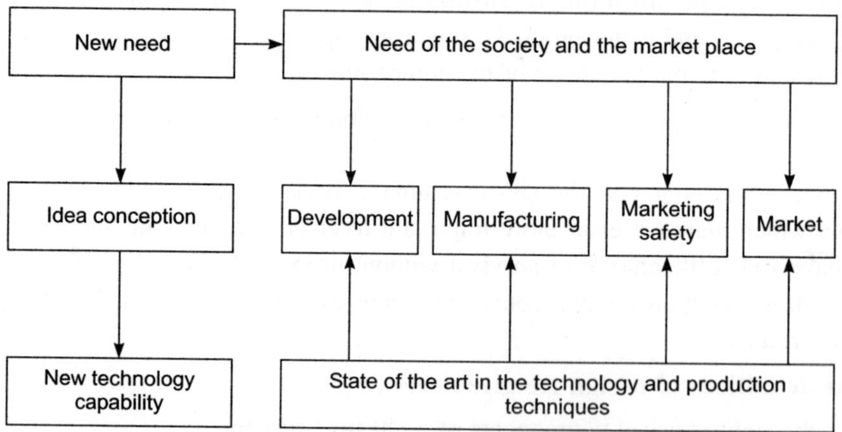


**Figure 1.2** Interactive model of innovation

Identification of various stages of *Product Development Cycle*, that starts at the initiating stage to the last phase of the product cycle and effective management of every stage is crucial for the successful implementation of product. These stages are diagrammatically shown in Figure 1.3.

The basic product idea or a trigger may be initiated from the various sources such as customer's need, from shop floor or new research and new technologies. Designers with their creative minds and abilities to look into the future can play a pivotal role in this triggering stage. The ultimate aim of product creation is to satisfy customer and induce a preferential attitude as compared to competitor's product.

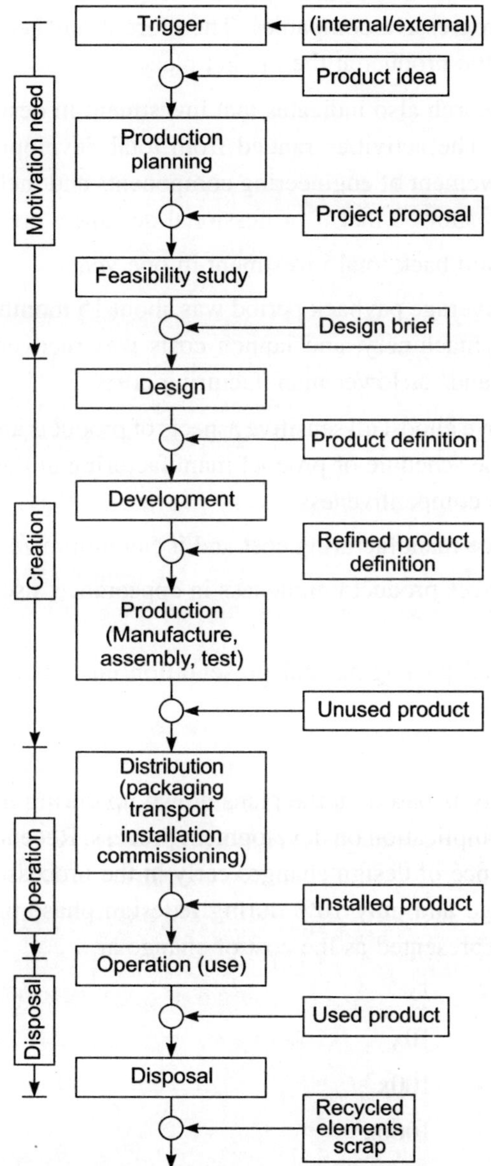


**Figure 1.3** Stages in the evaluation of a typical product

The product planning and feasibility stages help in developing an optimal design. Designer's role is to ensure the full analysis and understanding of the ultimate users' needs. Designer must work closely with other people from manufacture, assembly, testing, packing, transportation, and installation of product and even finally disposal of product at the end of its life cycle.

For any new product development process a huge investment is required for design, production, distribution and sales promotion without any confirmed commitment of buyers. For retaining the market position and to survive in today's competitive market, this risk activity of new product development is essential for progressive companies. Therefore it has become even more important to ensure commercial success of the product at the market place.

Interestingly the recent research also indicates that investment in design yields the highest returns for manufacturing companies. The activities ranged from total development of advanced electronic products, to incremental improvement of engineering components and include packaging, graphics and ergonomic designs. The results showed that if professional designers are involved:

- About 50% projects paid back total investment in one year.
- For other projects the average payback period was about 15 months. Full project cost including development, tooling, machinery, and launch costs was recovered through improved sales, higher profit margins and/ or lower manufacturing costs.

Design not only does improve almost all sensitive aspects of products and makes it more competitive, but this activity, also ensures the schedule of product manufacturing at the targeted cost. It contributes in three main areas to improve competitiveness:

(i) Cost: Design can reduce manufacturing cost and it can minimize recurring costs for the users.

(ii) Acceptability: It improves product uniqueness in appearance, user friendliness, reliability and safety.

(iii) Service: With improved packaging and presentation ensures safe delivery to the user and minimizes maintenance and repair.

The total production activity is based on the plans drawn up during the design phase. Any change during design phase has direct implication on development process. Recent successful product launches have demonstrated the importance of design changes early in the process. In Japan over 60% changes occur at product definition stage and only 10% during redesign phase prior to launch. The potential impact of any change can be represented as the cost of change at

| | | |
|---|---|---|
| Concept Stage | – | 1x |
| Design Stage | – | 10x |
| Tooling | – | 100x |
| Testing | – | 1000x |
| Post Release | – | 10000x |

## 1.2 RAPID PROTOTYPING

During last decade many new technologies were developed for *Rapid Prototyping (RP)* by which a physical three-dimensional plastic models are created, using CAD file data through special sintering, layering or deposition techniques, without building actual mould. This technology also known as *Solid Freeform Fabrication (SFF)* or *Layered Manufacturing (LM)* offers tremendous advantages in terms of saving the cost of prototyping, shorter lead times and improving total product quality.

Though currently most of these processes are very expensive, the use of RP will become much more acceptable with better, stronger and accurate models being built and will change its today's status of novelty. These processes are bridging the gap between design conceptualization and manufacturing.

An important phase of product development cycle is manufacture of production tooling. Characteristically all tools require long time for production and demand high investments. So many efforts have been put in to improve this phase to achieve savings in time and cost. Making these tools using conventional machine tools, CNC machines, EDM and so on, give very high dimensional accuracies but these machining operations are time consuming and costly. Rapid Tooling (RT) is an economical alternative to traditional tool manufacture.

Designer can play a very important role at the conceptualization stage of the product development that is critical for the success of a product in the competitive market. Product design is used as an effective weapon in the competitive market. Hence 'design' is now being discussed by all business enterprises. Designer's prime commitments to satisfy the users requirements, his needs, aspirations and safety considerations, become a pivotal activity for the design process. Designers' spatial creative abilities help him giving shape to the idea by effective 3D space utilization.

Feedback from those involved with the product after it leaves the factory — purchasers, operators, service engineers etc. — is vital for the designers providing essential information for the design of the second generation products.

Finally, and increasingly important as environmental factors influence profits more and more, designers need to be concerned about how the product is to be disposed of in order to simplify recycling or remanufacture and avoid pollution hazards.

## 1.3 CONCLUSIONS

During the course of product development, various decisions are taken that influence the cost and performance of the product. Any incorrect decision taken at one stage of the design process will get amplified by the subsequent stages. Therefore, early detection and correction of the design flaws cost less. Design has to take into account the features of subsequent activities in order to arrive at optimal decisions. DFX under Concurrent Engineering environment helps to achieve this goal.

CHAPTER 2

# Influence of Innovations on Product Development

## 2.1 IMPACT OF INNOVATIONS ON ECONOMY

International scenario is changing rapidly. Globalization and integration wave has swept the world trade and investment scene. These changes place number of challenges to all developing countries. Technological developments in the field of communication, production and marketing have been contributing to rapid growth in world trade and investment. Trade and financial flows have been increasing much faster than before, overtaking the growth of world output. Tremendous growth in trade and investments resulted the changes in the attitude and philosophy of world trade. Competitive spirit is growing among countries, each one is striving to be leading in this race.

**Table 2.1** Leading Asian countries in World Trade

| Country |
|---|
| China M.L. (Main Land) |
| China H.K. (Hong Kong) |
| India |
| Japan |
| Korea |
| Malaysia |
| Singapore |
| Taiwan |

No doubt that the globalization of trade and other changes are advantageous in the long run, but they have also added to the difficulties of developing countries. If developing countries want to be successful in this race, they have to make vigorous and consistent efforts. Being one of the developing countries, India is no exception to this. If we look at the performance of some of recently developed countries in total world export, emerging pattern is a clear indication how innovation is playing pivotal roll in developing economy of these countries (Table 2.1).

The percentage share of India in total world exports has diminished whereas the share of exports of Asian counties has miraculously increased. East Asian countries' success in increasing their share in world exports in short span of time is really a commendable task. This led to naming them as *Newly Industrialised Countries* or *Asian Tigers* (Table 2.2).

**Table 2.2** Selected Asian countries on the basis of their share in world exports

| Country |
| --- |
| China (M.L.) |
| China (H.K.) |
| India |
| Japan |
| Korea |
| Malaysia |
| Singapore |
| Taiwan |

According to World Development Report 1987 classification, trade policy can be divided into outward oriented and inward oriented. An outward oriented policy refers to trade and industrial policies that do not discriminate between production for domestic use and for export market, whereas inward oriented policy favours production for domestic market than the export market. Outward oriented policy promotes exports and thereby industrialization and economic development, but inward oriented policy encourages import substitution, protect domestic industries and encourage them to develop rapidly.

Objectives such as increasing market share, popularizing Indian brands in the world market, creating reputation of Indian industries in the world market did not appear in Indian trade policy. In contrast to this Japan, Taiwan, Korea and other East Asian countries followed outward oriented policy, chose export led growth and took aggressive approach. These countries since 50s tried to create their position in the world market by increasing their share in world export.

## 2.2 EXPORT COMPETITIVENESS

Export growth depends upon competitiveness of exports in the world market. Competitiveness is influenced by price and non-price factors. Indian goods lag behind in both ways, i.e., the prices of goods are not competitive to their rivals and quality of Indian goods is also not up to the mark.

The position of successful countries indicate that their policies gave enough attention to the quality of production, design aspects to make it more user friendly, cost structure and marketing efforts.

## 2.3 DESIGN AS A STRATEGY TO WIN INTERNATIONAL MARKET

The strategy used by Asian Tigers is product design and innovation to win international markets. As per this strategy, market can be captured if qualitative user-friendly products are innovated. These countries tried to find market niche and develop new products for exports (Table 2.3).

**Table 2.3** World competitiveness scoreboard

| Country | Ranking |
|---|---|
| Singapore | 2 |
| Hongkong | 7 |
| Japan | 16 |
| Taiwan | 18 |
| Malaysia | 27 |
| China | 29 |
| Thailand | 34 |
| Korea | 38 |
| India | 39 |

In the process of industrialization, these countries initially imitated the leading products, then improved them and finally developed new products indigenously and populárized their brands in the world market. According to this design strategy a comprehensive approach is adopted to incorporate all the good features in a product like:

- Better quality
- Reasonable pricing
- Attractive packaging
- User-friendliness.

These countries have concentrated their efforts in innovating new products for export markets (Table 2.4). Many innovations are based on a deeper level of customer and market understanding and go beyond what current customers say they need. These innovations create need for better performance once customers start using them.

**Table 2.4** Selected Economies based on expenditure in R & D

| Country |
|---|
| Japan |
| U.S. |
| South Korea |
| Canada |
| Singapore |
| China |

List of Asian countries selected based on the no. of patents registered is shown in Table 2.5.

**Table 2.5** Selected countries based on Patents registered.

| |
|---|
| • Japan |
| • Australia |
| • Korea |
| • China |
| • Hong Kong |

The product design approach has been approved and appreciated by even European countries to succeed in the world competition in the coming years. Further progress and improvements are to be made in design aspects. For this purpose our entire design system including design education should be revised and improved.

## 2.4 INNOVATION PROCESS

It calls for an intimate understanding of current customers and markets, potential new customers and markets, team and organization competencies and improvement opportunities, vision, values, and mission. Effective innovation depends on disciplined management systems and processes, that encourages a search for creative ways to do things better, different, or more effectively. Innovation leaders find a multitude of ways to understand and satisfy their customers. Many new products and services come from deeper understanding of the intended customers' problems and aspirations.

Customer's views are source of inspiration for developing new products and services.

Innovation process can be used to develop end products in a way, they have never been thought previously. These creative engineering devices are often derived by combination of known components, principles or devices in a unique, novel and non-conventional ways. Driving curiosity, will to explore unknown frontiers and wide imagination are playing important roles in this act of creativity. Albert

Einstein has rightly said "Imagination is more important than knowledge, for knowledge is limited, whereas imagination embraces the entire world – stimulating process, giving birth to evolution".

Almost every one of us is creative enough in early childhood, but very few can remain highly creative (about 2% in average at any given age) as they grow old. It is interesting to know, what goes wrong in the process of growing. The creativity of most of the individuals is inhibited by three different mental blocks, viz.,:

(i) Perceptual blocks

(ii) Cultural blocks

(iii) Emotional blocks.

### 2.4.1 Perceptual Blocks

These blocks are caused by not seeing the real problems or ignorance about what may be actually wrong. In the situation, these blocks tempt us to begin our problem solving activity without proper goal in mind. This tendency at the end, brings lot of frustration after looking at the most obvious solution for the problem.

These perceptual blocks may be classified as merely having a mental set or predisposition towards seeing the situation in a particular way only, no matter how closely or thoroughly we examine the problem. This attitude makes it difficult to define the problem in the beginning and the way the problem gets defined gives a specific direction to solution. This imposed blindness prevents looking at the solution from different area and its application to the given problem.

### 2.4.2 Cultural Blocks

Being members of a society, we are governed by the rules of behavior, and our thoughts and actions in a particular situation are reflections of customs and traditions. The society requires that an individual behaves in a certain way only because it is customary. While the creativity, on the other hand, calls for changes in the present ways, investigation of alternatives and if necessary proposes radical solutions. These blocks are first implanted at home and in pre-school years and are consolidated further by remarks like "Good Behavior, Good Grades etc".

People too curious and asking many questions are considered trouble some. By avoiding questioning we are attacking the very heart of creativity, i.e., killing curiosity. Only inquisitive minds can take new ways to make better things. Most of the times feasibility of solution and economic considerations cease our imagination.

### 2.4.3 Emotional Blocks

These blocks to creativity are within ourselves and we are afraid of making mistakes or making a fool of oneself. These over powering emotions like fear, love, hatred and anger can cause these emotional blocks.

Under the pretext of time pressure, we normally take the first idea occurring to our mind, as a good solution for the given problem and stop exploring further. Many a times the "really good idea" comes quite later as we warm up in this mental exercise. There is no magic formula to become more creative. Positive attitudes, self-analysis and the desire to improve are the most essential instruments for creativity.

The creativity should become like a personal trait by which, an individual sticks to the difficult problem, with an obsession and does not leave it until a good solution is struck. Most of the times creative discoveries happen when one

- lets his imagination soar to the sky
- select the alternative and then engineer it in the best possible way to bring it back to earth.

An American aluminum company coined a very precise word for this act "*Imagineering*". Only after casting off the traditional thinking, building immunity towards negative thinking and accepting boldly the challenges to achieve "impossible" one can succeed in the act of "imagineering".

A larger pool of less significant ideas allows us to select the most effective idea, as the mathematical probability of finding a really creative solution increases manifold. The process through which one can make a long, exhaustive list of alternatives is called *ideation*.

The important source of building a bank of ideas is through the phenomenon of association of ideas. In this process, one imagines something else when seeing, hearing, tasting, feeling or smelling the existent and builds one idea based on another. The skill of coming up with new ideas can be increased with constant practice and close awareness of nature, man made objects and day-to-day occurrences of the events. One can analyze the problem most critically by asking some of the following questions:

- What is wrong with present product? An exhaustive list of all the lacuna should made.
- How can this be improved? All those bad features listed above are to be improved at the same time considering feasibility.
- Can this be used for other applications? List all other possible users for the present product and also, those applications with some modifications in the present product.
- How an appearance of product be improved? Various options like colour, form, textures, graphics, packaging, the material for the casing are to be listed, and the economics of change.
- How the present product can be made more convenient to use and safer to operate? Make it easier to work with, easier to carry around, easier to dispose, faster in operation, automatic, silent, safer to use, safer to children, non toxic, nonflammable, stain proof and so on.

Another approach can be taken in Ideation is by transformation of existing product:

- *Modify* – changes trims (shapes), color, form, material—
- *Magnify* – make it stronger, thicker, larger, longer—
- *"Minify"* – make smaller, shorter, lighter, narrower
- *Reverse* – try to make it up side down, around, operating in opposite way.

- *Rearrange* – interchange components, put it in new order—
- *Replace* – plastics for metal, light colour instead of dark, round form instead of square—
- *Combine* – Components, principles of working, methods of assembly various functions in one
- *Adapt* – New application for old product, association of something else, copying old principles in newer context.

## 2.5 CONCLUSIONS

Various tools described so far and many other like brain storming, idea matrix, role playing, synectics and bionics are to be used for making an exhaustive list of alternatives, which enables one for "seeing it different ways" the existing situation, so that many new avenues are thrown open for new product development.

CHAPTER 3

# Rapid Product Development — An Overview

## 3.1 INTRODUCTION

In the buyers' market of today, ability to compress *time to market* and *agility* to respond to changing customer needs decide the survival of manufacturing organizations. *Rapid Product Development (RPD)* enhances these capabilities of the organizations and hence their competitive advantage.

RPD is the synergic integration of the various time compression technologies — which may be virtual or physical in nature. Time compression can be achieved by doing the various design, testing and manufacturing activities in the virtual world as far as possible. When this becomes no longer possible, one can shift to the physical world where the aim is to do them rapidly. This is the philosophy behind RPD.

The following are the three pillars of RPD (Figure 3.1):

(i) Virtual prototyping and testing technologies,

(ii) Physical prototyping and rapid manufacturing technologies

(iii) Synergic integration technologies.

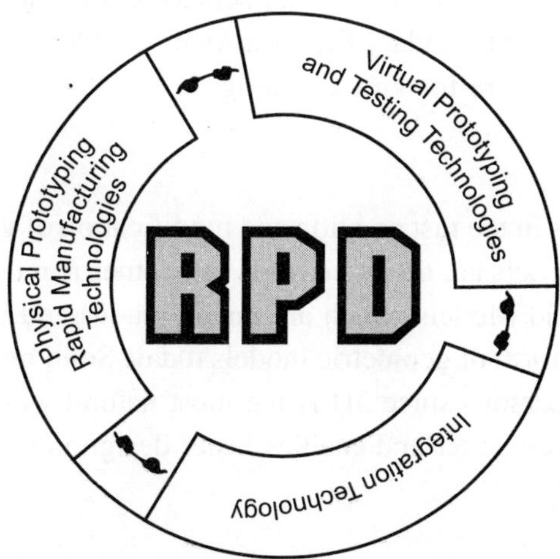


**Figure 3.1** Three pillars of Rapid Product Development

## 3.2 VIRTUAL PROTOTYPING AND TESTING TECHNOLOGIES

When an activity of the product development cycle is carried out in a virtual or soft environment, the primary benefit is the time saving and hence cost saving. There are also few other important benefits as follows:

- *Minimum risk to human life*: VR (Virtual Reality) based flight simulators help the pilot prepare himself well before the fighter plane is ready for flying.
- *Upward or downward scalability of time in virtual world*: The pace with which an activity takes place can be increased or decreased as required if it happens in a virtual world. It is possible to simulate creep of components in a short time to assess their fatigue life which otherwise would take several hours when tested physically. On the other hand, one can analyze the trajectory of a missile leisurely at a chosen pace which actually may be over in a few minutes.
- *Less fatigue to the personnel:* Monotonous steps in various activities are automated. For instance, drawings with various types of views and dimensioning can be automatically generated from the solid models with a few commands.

The following are the technologies for prototyping and testing in the virtual world:

- Geometric Modelling
- Reverse Engineering
- Virtual and Augmented Reality
- Finite Element Analysis
- Boundary Element Analysis
- Motion Analysis
- Tolerance Analysis and Synthesis
- "Design for X" etc.

These are essentially *Computer-Aided Design & Engineering (CAD & CAE)* tools. Geometric Modelling, Reverse Engineering and Virtual and Augmented Reality are used for virtual prototyping and tools such as Finite Element and Boundary Element Analysis, Motion Analysis, Tolerance Analysis and Synthesis, "Design for X" etc. are for virtual testing.

### Geometric Modelling

*Geometric Modelling (GM)* helps in the fast creation and modification of virtual models of the products. *Wire-frame*, *surface* and *solid* models are the three types of geometric models of increasing complexity and richness of information. Solid Models which are unambiguous and in complete representation of the products is the most popular form of geometric models today. Solid models can be created elegantly and easily even by less skilled persons since 3D is the most natural way of designing. Feature-based solid modelling makes things even better and enables faster design iterations.

### Reverse Engineering

*Reverse Engineering (RE)* is used to create geometric models of physical objects. This is an important step in design iteration, especially of form design. RE also comes handy when it is required to reproduce existing objects, especially old or military equipment where the supplier no longer exists or is unwilling to help. RE in conjunction with RP&T can influence in a big way prosthesis is practiced although it is yet to be exploited in our country.

### Virtual Reality and Augmented Reality

This helps in better visualization as well as the evaluation of the design. In VR the objects exist only in the virtual world whereas they exist both in virtual and physical form in AR. These are useful for assembly simulation, ergonomic analysis, simulation of surgical procedures etc. AR has been used successfully in carrying out minute laproscopic operations.

### Finite Element and Boundary Element Analysis

*Finite Element Analysis (FEA)* and *Boundary Element Analysis (BEA)* are tools for virtual testing. These can be used to assess the stress levels of components under various types of static and dynamic loads. These loads may be purely mechanical or a combination of fluid pressure and thermal loads. These are also helpful in analyzing fluid and thermal flow problems. Although BEA was developed earlier, its complex integral mathematics was responsible for its trailing behind FEA in popularity. FEA is used for various kinds analyses thanks to the availability of several commercial FEA packages. With the advent of symbolic algebra, BEA is becoming popular in certain niche areas.

### Motion Analysis

Several software tools are available that can do motion analysis of assemblies to find out their working volume, accelerations at various positions etc. Interference or collision detection is also one of the most important uses of motion analysis.

### "Design for X"

*Design for X (DFX)* is an important technology of RPD. *Design for Manufacture & Assembly (DFM&A)* came first. Subsequently, these concepts could be extended to various other activities such as dissassembly, repair, recycling etc. DFX essentially is a set of CAE techniques that help in evaluating the design at a very early stage. DFX and Concurrent Engineering together help in early detection of design flaws. The later an error is detected, the more it costs to correct. Based on the result of these evaluations, design modifications are made.

## 3.3 PHYSICAL PROTOTYPING AND RAPID MANUFACTURING TECHNOLOGIES

In order to compress product development cycle time, as mentioned earlier, activities are carried out in virtual world as far as possible and when this is no longer possible, they are done rapidly in the physical world. For achieving the speed, following compromises on material, quality and cost have been found acceptable:

*Material*: Most RP machines make prototypes out of soft non-metallic materials, that too of proprietary nature. These prototypes are not isotropic. In spite of these limitations, these prototypes help in rapid assessment of form, fit and functional performances of the design, to varying extents.

*Quality*: The accuracy and surface finish of RP parts are nowhere near those of the machined parts. However, such prototypes enable several design iterations within the given time.

*Cost*: The cost of a RP part may be much higher than the one made by a pattern maker or by conventional or NC machining. However, RP can make it faster. Therefore, where time-to-market matters, money spent in RP parts will lead to net profit although it may appear suboptimal initially.

The following are the technologies for physical prototyping and rapid manufacturing:

- Computer Numerical Control
- Robotics
- Computer-Aided Process Planning
- Rapid Prototyping & Tooling etc.

These are essentially *Computer-Aided Manufacturing (CAM)* technologies.

### Computer Numerical Control & Robotics

*Computer Numerical Control (CNC)* machines and *robots* are automatic machines working on similar principles of drives and controls. They primarily differ in their kinematic configurations and hence differ in terms of accuracy, rigidity and dexterity. In fact, a CNC machine can be loosely called as a Cartesian robot. The other subtle differences between CNC machines and robots are the extent of sensory perception and the method of programming. CNC machines and robots are known as machines of *flexible automation* type since change over from one product to the other is achieved simply by changing the control programs. CNC machines are used primarily for component manufacture whereas robots are mostly used for material handling, assembly, arc welding, deburring, spray painting etc.

### Computer-Assisted Process Planning

*Process Planning (PP)* converts the design information available in the form of geometry, tolerances and material specifications into manufacturing information in the form of motions on machine slides and other process parameters. Therefore, this is considered a bridge between design and manufacture. When use of computers is employed for process planning, it is known as *Computer-Assisted Process Planning (CAPP)*. CAPP is considered a bridge between CAD and CAM. In a conventional manufacturing environment, CAPP will generate a process plan that indicates the sequence of operations to realize a component, sub-assembly or assembly along with the details of the process, tooling requirements, set up and unit times etc. When the acronym *CAPP* is expanded as *Computer-Assisted Part Programming*, the outcome is not a conventional process plan but the NC programs to carry out the operations on a CNC machine.

There are primarily two types of CAPP systems:

(i) Variant process planning

(ii) Generative process planning.

In variant process planning, a process plan existing in the company for a similar component is retrieved and suitably modified so that the new component can be made using it. This is a semi-automatic method which relies on *Group Technology* (*GT*) to classify and retrieve similar components. Generative process planning is fully automatic. It interacts with the various databases (DBs) of machine tools, cutting tools, production tools etc. and knowledge bases (KBs) consisting of heuristics and thumb rules extracted from expert process planners and generates a conducive process plan. The preferred geometric input to automatic process planning is *feature-based.* If a conventional B-Rep model only is available as input, then a preprocessing step called *feature extraction* is used. Unfortunately, only limited success has been achieved in generative process planning that too for simple geometries such as axi-symmetric and prismatic features.

### Rapid Prototyping & Tooling

The benefits of both CNC machines and robots can be fully realized only when their fool-proof tool path programmes for any operation can be generated automatically. Due to the limited success in CAPP systems, this is still not possible and hence a skilled NC programmer has to generate the control programmes. An alternative to CNC is *Rapid Prototyping & Tooling (RP&T)* technology wherein the required object can be manufactured from its CAD model automatically without the use of any tools. Although RP&T could make the prototypes only out of proprietary non-metallic material and the accuracy and surface finish are not adequate, direct and indirect methods are emerging to overcome these limitations. Hybrid technologies which exploit the best features of CNC and RP are also emerging. These efforts will shortly enable making objects rapidly from their CAD models automatically out of the required material with the required quality.

The huge investments on RP machines as well as the high cost of these prototypes can be amortized better if the same technology is able to produce tools through direct or indirect routes. It may be noted that processes like *Selective Laser Sintering* and *3D Printing* are capable of producing prototypes from a wide variety of materials right from wax or polystyrene to metals. When the same technology can be extended for manufacturing parts in small batches, then it will be called Rapid Manufacturing.

## 3.4 SYNERGIC INTEGRATION TECHNOLOGIES

Product Development is a network of several activities. The time taken by an activity may be constant or variable. For instance, an automatic operation may have a fairly constant time whereas the cycle time of a manual operation may have wide variations. Some activities can happen only if a few others are complete whereas there could be many sets of activities in the network which can progress together. Therefore, RPD cannot be fully realized if these factors are not accounted for and managed properly. In other words, we need certain managerial tools that ensure synergy among the various activities so that the benefits achieved at individual activities are carried forward without significant attenuation till the end.

The technologies used for the synergic integration of the activities of product development cycle as well as product life cycle are:

- Concurrent Engineering
- Product Data Management

- E-Commerce
- Computer-Integrated Manufacturing etc.

### Concurrent Engineering

An important technique to cut down product development time is to carry out as many activities in parallel as possible. This is enabled by a technology called *Concurrent Engineering (CE)*. The reduction in time happens not only by parallel processing but also by an early detection and correction of error due to a feedback from related groups. Therefore, sometimes, *Concurrent Engineering* is also called *Collaborative Engineering*.

### Product Data management

Since the customers vary in terms of taste, buying power etc., the same organization has to make several models of the product each with its own variations and options. Furthermore, they have to release new products year after year while they are still committed to support earlier products for a certain period. All these make manufacturing very complex requiring sophisticated systems called *Product Data Management (PDM)* systems to keep track of various activities and entities.

### E-Commerce

The influence of Internet and intranet to manage global organizations hardly needs emphasis. More and more software get web-enabled to exploit these facilities. All these are called *E-Commerce*.

### Computer-Integrated Manufacturing

*Computer-Integrated Manufacturing (CIM)* is an embodiment of all the above integration technologies. Another popular term used with similar meaning is *Intelligent Manufacturing System (IMS)*.

There are three prerequisites for CIM/ IMS implementation in an organization:

(i) All activities must happen on computers.

(ii) There must be reliable connectivity among these computers through LAN/ Internet/ Intranet.

(iii) There must be well-established communication protocols among the computer systems. Furthermore, there shall also be a reliable security system to monitor and control the data that may be distributed across the network.

RPD leverages the ability of an organization to introduce new products and versions rapidly to meet the fast-changing customer needs. The one reaching the market first can own the market.

# Part II

## VIRTUAL PROTOTYPING AND TESTING

# CHAPTER 4

# Geometric Modelling

## 4.1 INTRODUCTION

All physical objects in the world are 3D in nature. However, in order to convey one's own ideas of a 3D object in his mind, it is not always possible and often not practicable to show by creating the physical shape in materials such as clay. Therefore, there was always an urge to depict 3D shapes as realistically as possible without actually creating their physical models. Since paper happens to be the easiest medium, techniques were developed by which it is possible to represent 3D objects by means of one or more 2D sketches or views or pictures on it. Some of these techniques used in depicting engineering objects are orthographic projections (standard views, auxiliary views and sectional views), axanometric projections (trimetric, dimetric and isometric views), oblique projections and perspective projections. Orthographic projections can hold a lot of manufacturing details of dimensions, tolerances, material specifications, BOM etc. in text form and hence they find use in shop floor while the others are pictorial in nature and are used for better visualization. When very huge objects such as bridges, buildings etc. are to be depicted pictorially, perspective views are used. These views are 2D in nature and hence require some amount of imagination on the part of the person to interpret them. The collection of these techniques used to represent 3D object in terms of 2D views is called *Engineering Graphics*. When use of computers is employed for the above Engineering Graphic work, Engineering Graphics comes to be known as *Computer Graphics*. (It may be noted that this definition of Computer Graphics no longer holds good since today Computer Graphics encompasses all techniques that are used for representation of objects, interaction with object models, preparation and presentation of the views).

Since initially only 2D details could be depicted in a computer, Computer Graphics was used only for drafting. Therefore, CAD at that time was almost synonymous with "Computer-Aided Drafting". This phase of design can also be called *View-Based Design*. This resulted in the following benefits:

- High speed
- High accuracy
- High quality of design
- Less fatigue to the designer.

Subsequently, techniques to represent 3D geometry in the computer were developed. This led to the possibility of representing objects in 3D. Therefore, the use of these models were not limited to only drafting; they could be directly used as input to various downstream functions of Computer-Aided Engineering and Manufacturing (CAE & CAM) such as Finite Element Analysis, NC cutter path generation, tool design, process planning etc. This phase of CAD can be called *Model-Based Design*. The impact of 3D modeling was so much that it could eliminate the need for drafting in many organizations leading to "paperless manufacture".

*Computer Graphics*, *Geometric Modelling* and *Computational Geometry* are often used interchangeably although there are subtle differences. Geometric Modelling refers to representing, modifying and interacting with virtual objects on computers. Computational Geometry provides the necessary theory for this. In other words, if Geometric Modelling is analogous to a computer programme implementation, Computational Geometry is analogous to its algorithm. Computer Graphics is a superset of both.

## 4.2 TYPES OF GEOMETRIC MODELS

In Geometric Modelling, three types of 3D representation schemes are used:

(i) Wire-frame representation

(ii) Surface representation

(iii) Solid representation.

In order to understand these representations, it is important to know the meaning of two terms, viz., *geometry* and *topology* since these two details are required to define an object. Geometry describes the location and size of entities whereas topology tells how the entities are connected among themselves. In other words, topology of an object describes how faces are bounded by edges, how edges are shared by faces, how vertices are shared by faces, how vertices are shared by edges and so on. For example, coordinates of a point, the radius of a circle etc. will be called geometry whereas the information as to whether a given arc or a corner is concave or convex will be called topology. Figure 4.1 has two shapes which have the same geometry but different topology.

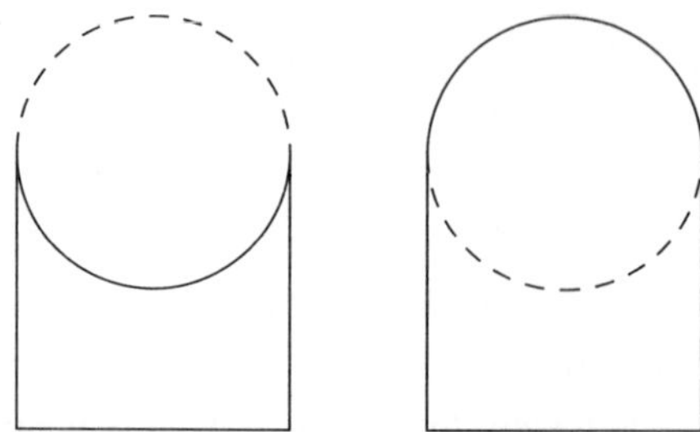

**Figure 4.1** Two objects of same geometry but different topology

Wire-frame, surface and solid representations are distinguished by the extent of topological information they have. Table 4.1 shows the geometric and topological information stored typically by these representations. From this table, it is obvious that the former model is a subset of the latter ones in that order. Because of this characteristic, all solid modelling systems make use of the wire-frame and surface models for quick intermediate displays.

**Table 4.1** Geometric and topological information stored in various types of geometric models

| Geometric Model | Geometry | Topology |
|---|---|---|
| Wire-frame model | Vertices | Edges (Connectivity among vertices) |
| Surface model | Vertices | Edges, Faces (Connectivity among edges) |
| Solid model | Vertices | Edges, Faces, Solids (Connectivity among faces) |

### 4.2.1 Wire-Frame Models

A wire-frame model is analogous to obtaining the shape of an object by welding the wires representing the edges together. It is the simplest of 3D representation schemes and hence the fastest. Therefore, surface and solid modelers make use of it for generating quick displays. However, the usefulness of wire-frame models is limited because of the following drawbacks:

- A wire-frame model is ambiguous since the same model can be interpreted in several ways. This is illustrated in Figure 4.2 where a block is shown in three different interpretations.

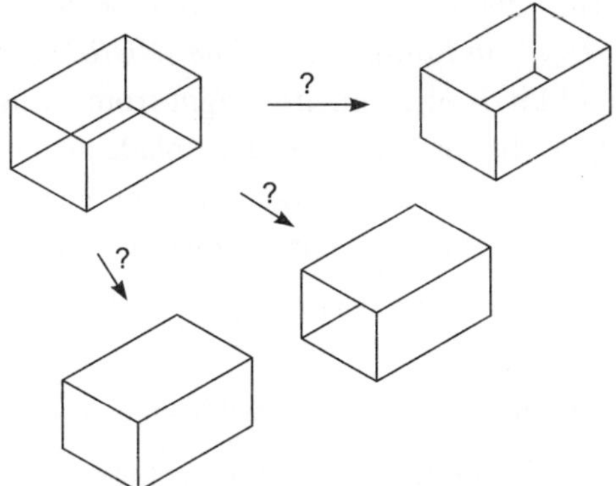


**Figure 4.2** Different interpretations of a box in wire-frame modeling

- It is possible to create nonsense objects in wire-frame representation. The examples shown in Figure 4.3 illustrate this.

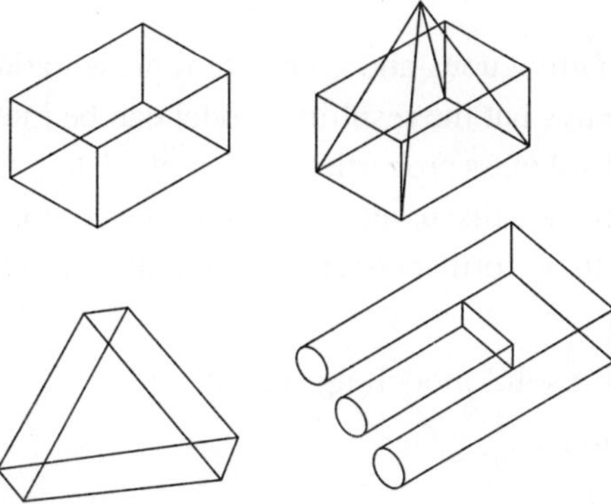

**Figure 4.3** Some nonsense objects possible in wire-frame and surface models

- The topological information available in a wire-frame representation is inadequate to create hidden line removed views. Therefore, the object looks messy making interpretation difficult.
- Calculation of mass properties such as surface area, volume, mass, center of gravity, moment of inertia etc. is not possible for a wire-frame model since only edges are available.

### 4.2.2 Surface Models

A surface model can be of two types, viz.,

(i) Faceted or tessellated (small flat pieces joined) surface model

(ii) Exact surface model.

A faceted surface model is analogous to obtaining the shape by pasting together the cardboard pieces that represent the faces. Since it is simple and a subset of any solid modeler, it is often used for generating quick hidden line removed displays. Till recently, it was possible to use only simpler shapes such as block, wedge, sphere, cylinder, cone and torus in solid modelers. Therefore, for applications that require freeform surface modeling as in styling and applications where accuracy is very important as in NC path generation, exact surface modelers were used in these areas. The exact surface modelers most commonly make use *of Non-Uniform Rational B-Spline (NURBS)* data structure. The advantage of NURBS is that it can represent all kinds of geometries right from straight line to more complicated higher order free-form surfaces such as the surface of a turbine blade using a unified data type. However, solid modelers are replacing such surface modelers today as they can also handle with greater ease freeform surfaces now. Some of the limitations of surface modelers are:

- Faceted surface models are inaccurate. If accuracy is to be improved, the size of the model becomes huge.
- Nonsense objects given in Figure 4.3 can be possible in a surface model also.
- Stitching of edges exactly is difficult, i.e., the models may not be "water tight" or may have dangling patches.
- On a pure surface model, only surface area can be calculated and other mass properties are meaningless since details inside these surfaces are not defined.

### 4.2.3 Solid Models

A solid model is said to be an unambiguous and complete representation of a physical object. A solid model can be created in several ways but the resulting model can be interpreted only in one way. Since the inside of a solid is well defined, all mass properties can be calculated. Although the solid representation is the costliest of the geometric models discussed so far in terms compute space (RAM and disc space) and time, the rapidly falling price to performance ratio of computer hardware and software has overcome this difficulty.

Some of the requirements of a solid modelling scheme are:

- Accuracy of representation
- Space complexity
- Time complexity

- Ability to handle self-intersecting objects
- Ease of creation
- Ease of modification
- Complexity of the shapes that can be represented.

Over the period, several solid model representations have been developed. Some of them are:

- Constructive Solid Geometry models
- Boundary Representation models
- Spatial decomposition models
- Feature-based models.

Each one of them has advantages and limitations and accordingly they also find specific applications. Among these, CSG and feature-based models are called unevaluated models and the other two are in evaluated form.

## 4.3 TYPES OF SOLID MODELS

Several solid representation schemes have been tried. These have user-friendliness and system friendliness to varying extents. In general, a scheme that has good user-friendliness has poor system friendliness and vice versa. Four of the most important ones among them are discussed here. It may be noted that a typical CAD package makes use of two solid representation schemes to obtain satisfactory levels of user-friendliness and system friendliness.

### 4.3.1 Constructive Solid Geometry Models

*Constructive Solid Geometry (CSG)* is the first solid modeling type developed at University of Rochester, U.S.A. Certain geometries such as blocks and cylinders as shown in Figure 4.4 are very familiar to the users

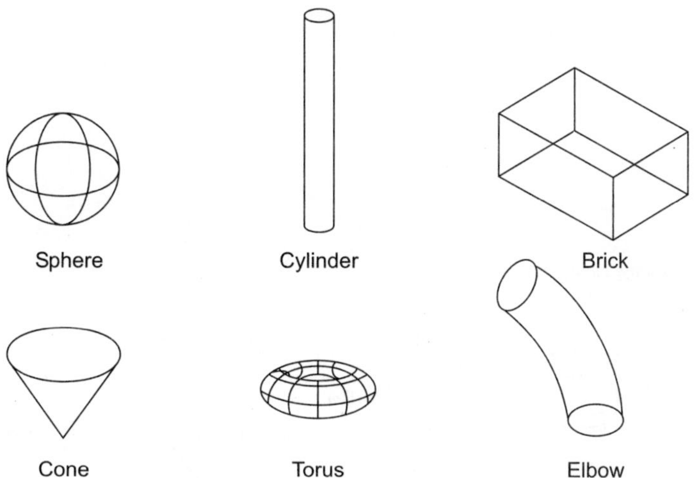


**Figure 4.4** Primitives used in CSG models

and the concept in modelling is to use these simpler known geometries called *primitives* to construct more complicated objects. The process of construction makes use of three operators called *Boolean* operators. These operators are:

(i) Union
(ii) Subtraction
(iii) Intersection.

Just as one would add and subtract numbers, in CSG modeling, the geometries are manipulated using these operators. It is possible to represent the primitives and Boolean operators used in constructing any object in the form of a tree called *CSG tree*. The leaves of the CSG tree of any object will be the primitives used and the branches will be the Boolean operators. The root of the tree is the constructed object itself. A few other operators are also used to edit the object to get features such as filleting and chamfering and transformations. However, they are not shown in CSG tree. The example of the CSG tree of a tappet valve is shown in Figure 4.5.

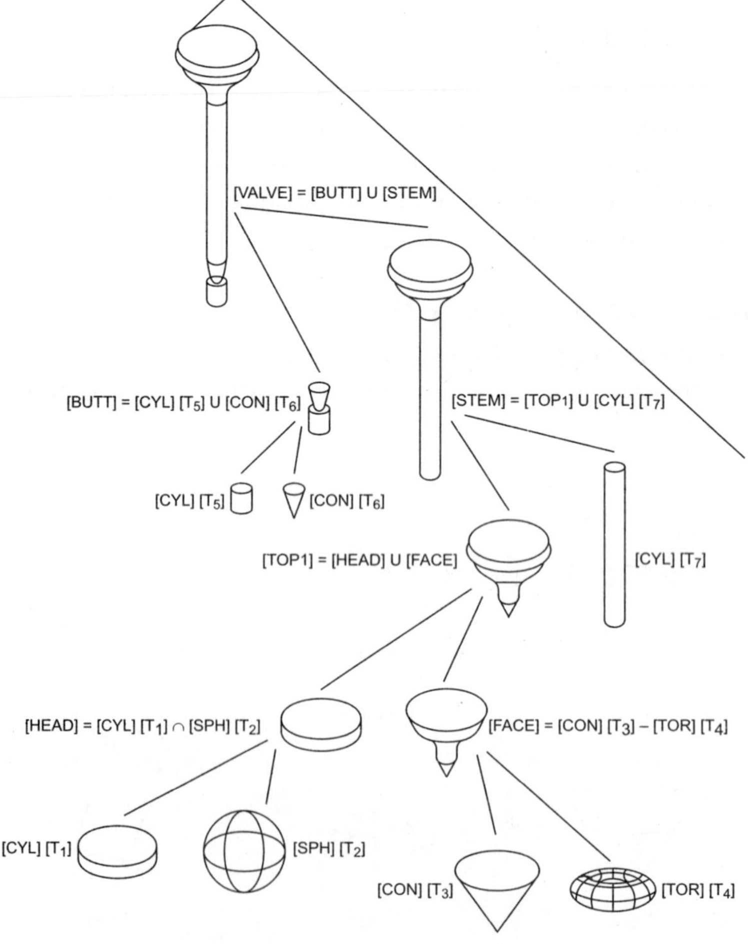


**Figure 4.5** CSG tree of a tappet valve

*Advantages*

- Simple to understand and use
- Low memory requirements.

*Limitations*

- CSG is an unevaluated or raw model. Therefore, for any operation such as display or mass property calculations, it takes more time since all the necessary faces and edges are to be calculated every time.
- Data accession becomes more difficult as the complexity of the object increases.

### 4.3.2 Boundary Representation Models

*Boundary Representation (B-Rep)* is a very powerful solid modelling scheme that makes use of the concept of half spaces. Any surface that can be mathematically represented, be it a plane, sphere etc., divides the universe into two halves. Any object boundary is constructed out of such surfaces each of which divide the universe into two halves. While the equations of these surfaces are equalities, the corresponding half spaces are represented by the respective inequalities. One of these inequalities (or half space) represents the side on which material is present. A B-Rep model is obtained as an intersection of all inequalities of the object boundaries corresponding to the material side. B-Rep models are generally represented by a *winged data structure* with adequate redundancy for the benefit of speed. Figure 4.6 shows the details of an object in B-Rep. modelling.

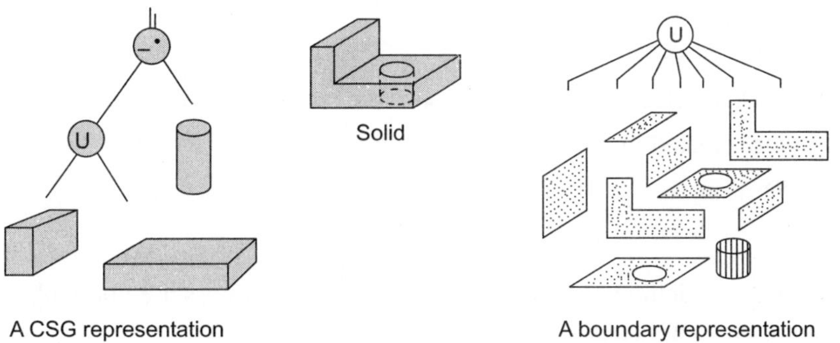


**Figure 4.6** B-Rep vs. CSG

*Advantages*

- Algorithms work very fast due to its presence in evaluated form and data redundancy.

*Limitations*

- B-Rep models occupy large space.
- They are difficult for a user to create since he has to calculate the intersections of various surfaces. In other words, they are less user-friendly.

- B-Rep models have large amount of data redundancy.
- The topology of the object may be disturbed during manipulations if adequate care is not taken leading to nonsense objects.

### 4.3.3 Feature-Based Models

The word *feature* refers here to the manufacturing features such as holes, slots, bosses etc. The dimensions are always stored as parameters inside this model. If a hole is to be called a hole, the full circular cross-section of it must be within the object lest it becomes a slot. Therefore, the parameters of any feature cannot take any value; these values have to be within certain constraints. These constraints may be geometric relations such as perpendicularity, parallelism, concentricity etc. or they can be dimensional values such as lengths and angles. Therefore the term *Feature-Based Model* actually refers to *Feature-Based Constrained Parametric Model.* By changing the values of the parameters within the constraints, one can get a variety of shapes and sizes. Therefore, one can actually design a family of parts rather than a part in the same time. With proper planning by the user, this can lead to dramatic improvement in productivity.

**Procedure for Constructing a Feature-Based Model**

- Choose a sketch plane.
- Sketch a rough 2D sketch on the sketch plane. This gives roughly the topology.
- Restrict this sketch in the following three levels:
  - Use the rules defined internally
  - Add more relational constraints
  - Add dimensional constraints.
- Convert this 2D sketch into a 3D feature such as extrusion, sweep, cutout, slot, hole etc.
- The first feature created in this manner is called *base feature*. Use the above steps to create all other features.

***Advantages***

- In feature-based modelling, unlike CSG, the Boolean operators are not explicit. They depend on the feature characteristics.
- By changing a few parameters, the object can be changed unambiguously since all the dimensions are related to each other by these parameters. In this way, a family of parts can be designed with the same effort required to design a part in CSG modelling.
- It enforces just dimensioning.
- It can do automatic dimensioning.
- Due to the presence of constraints, even if the dimensions are changed, the topological relations are preserved. For instance, a through hole remains as a through hole even if the thickness of the plate is increased.
- Creation of 2D sketches and their conversion into 3D using familiar features makes this approach more elegant and natural.

### *Limitations*

- Comparatively higher intelligence is required for the users of feature-based modelling. Solving constraints requires considerable amount of geometric acumen.
- User should plan well in deciding the right parameters to exploit the benefits of this group design philosophy.

## 4.3.4 Spatial Decomposition Models

Spatial decomposition models are of two types, viz.,

(i) Exhaustive enumeration models

(ii) Hierarchical Space Decomposition.

The exhaustive enumeration techniques represent the object in terms of equally sized cubes or spheres or rods of equal diameter (See Figure 4.7). This technique, although very simple, is of less

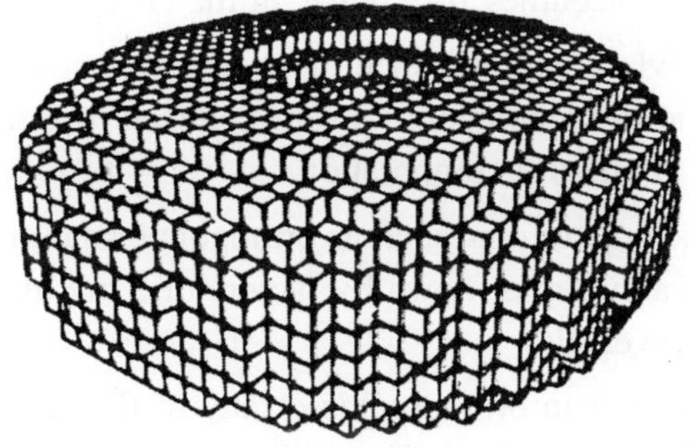

**Figure 4.7** Solid representation of a torus by exhaustive enumeration

practical use since higher resolution is demanded in many applications, it becomes very huge. Therefore, it is limited to a few niche applications. To overcome this difficulty of high space complexity of exhaustive enumeration scheme, *Hierarchical Space Decomposition (HSD)* models were introduced. It may be of many types such as octree, bintree, polytree etc. In HSD scheme, as shown in Figure 4.8, the

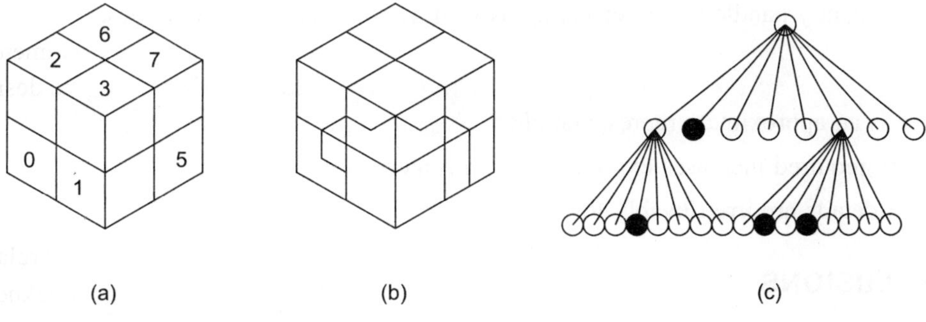


**Figure 4.8** Octree representation of a part (a) Universe, (b) Component, (c) Octree

universe (a cube that contains the object) is subdivided into 8 cubes by halving it across the principal directions. Each of these 8 cubes is categorized as *FULL* if it lies fully inside the object, *EMPTY* if it does not touch the object at all and *PARTIAL* otherwise. While EMPTY and FULL nodes become the leaves of the tree called *Octree*, the root will be the object itself. The PARTIAL cubes are the intermediate nodes of this tree and hence are further subdivided and categorized in the same manner recursively till the required resolution is reached. Because of the nature of the HSD models, they are commonly used in places where the number of primitives and Boolean operations are vast as in the case of NC simulation. They are also ideal to represent ill-defined lump geometries such as terrains, tomography etc. It may be noted that the size of the octree of the object depends on only its surface area for any given resolution and not on the number of primitives and Boolean operations.

***Advantages***

- Memory required by HSD models is independent of the number of primitives and operations. For a given resolution, memory required depends only on the surface area of the object.
- One can choose any desired accuracy (of course, at the cost of memory and speed).
- Boolean operations are trivially simple since these operations require only tree traversals. In other words, the geometric Boolean operations are reduced to Boolean operations of the bits. It requires absolutely no geometric computation.
- Many other frequently used operations such as orthogonal rotations, translations of objects in one octant to the other, scaling up or down by factors of 2 are also trivial since these too are achieved by simple tree traversals.
- Similarly, display of the object in orthographic or any of the eight isometric projections also can be achieved by simple tree traversal. These views can be quasi-rendered using three basic colours with hidden line removal. Since absolutely no geometric computation is involved and all computations are in unsigned integer space, the displays are extremely fast.
- All the computations take place in binary or unsigned integer modes. This feature minimizes space and time complexity of the algorithms.
- It inherently lends itself to parallelization. Since PCs are available today with multiple processors at affordable prices, this feature will be a major advantage.
- It can elegantly handle self-intersecting as well as non-manifold geometries.

***Limitations***

- Octree is an approximate representation.
- Memory required increases exponentially with increase in resolution.
- Feature details are irretrievably lost.

## 4.4 CONCLUSIONS

Geometric modelling is the first step in any product development. Solid modelling has become synonymous to geometric modelling. Only some styling applications still make use of surface models.

This too is mainly because the stylist does not care to create the complete topology required to define solid as he has to evaluate several alternatives and often a surface model is good enough for him.

From the earlier discussions, it is clear that no single solid model scheme meets all the requirements. Modelers such as CSG and Feature-based modelers are very user-friendly but they are computationally inefficient, i.e., they are not system-friendly. On the other hand, B-Rep and HSD schemes are highly system-friendly but are poor in user-friendliness. Therefore, any practical solid modelling system today invariably makes use of dual or multiple representations — one being user-friendly and the other being system-friendly. This is illustrated in Figure 4.9. Some examples are given in Table 4.2. As soon as the user updates in user-friendly form, it is immediately converted into the system-friendly form. However, it is generally not required to convert the system-friendly form into the user-friendly form for general modelling applications.

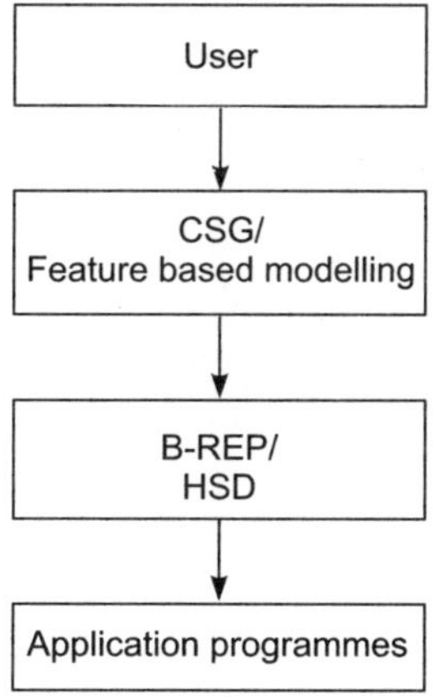


**Figure 4.9** A practical solid modelling system

**Table 4.2** Solid representations used by some popular geometric software packages

| Software | User-friendly modelling | System-friendly modelling |
|---|---|---|
| AutoCAD, I-DEAS, Unigraphics | CSG & FBM | B-Rep. |
| Pro/Engineer, SolidWorks, SolidEdge | Feature-based | B-Rep. |
| Materialise's Magic, Deskarte's Designer | CSG & STL input | Polyhedral B-Rep |
| Software for GIS, medicine, NC simulation | Images, CT and MRI scans, DSG (Destructive CSG) | Voxel or HSD |

Most popular CAD packages make use of higher level geometric definitions such as Bezier and B-spline curves and surfaces. Very powerful solid modelling libraries or kernels such as *Parasolid*, *ACIS* and *CASCADE* are available today. In order to speed up CAD packages, developers base their development on one of these kernels. More recent technologies such as RP and VR make use of polyhedral (i.e., faceted or tessellated) solid models. They accept solid geometry input as a set of boundary triangles. Even a very complex model with freeform features can be defined using STL format although these files could be very large — as much as 100 MB in binary format. Object definition is in the form of a set of triangular facets. Although engineers are not happy with STL format, they live with it as no better format could be developed for over a decade. Even graphic libraries such as OpenGL use triangular faces for rendering. This necessitates the need for triangular solid modelling kernels to handle huge files efficiently. Today it does not exist commercially and hence developers in RP and VR have to develop their own libraries. Such polyhedral kernels will help reduce geometric software development time.

CHAPTER 5

# Reverse Engineering

## 5.1 INTRODUCTION

*Reverse Engineering (RE)* is the process of creating a mathematical representation or CAD model of an object from its physical form. RE is necessary when

- A part is first modelled in clay, wood or foam and need details to be transferred into CAD
- Only 2D drawings or master models of physical tooling exist
- A competitor's product needs to be analyzed
- A change in a physical part or tool must be captured in CAD
- Final parts have to be verified against the original CAD design.

The process consists of first scanning the object to create *clouds of points (CoPs)* that represent the skin of the object. These points on the boundary of the object are used to create the surface model of the object using software such as *Surfacer*. If enough topological details are available, the surface model can be converted into a solid model. Otherwise, user can provide the necessary information to convert it into a solid model. The solid model thus created can be then used for analysis, documentation, and NC cutter path generation or for prototyping. RE essentially consists of two steps (Figure 5.1):

(i) Acquiring point data

(ii) Constructing 3D model.

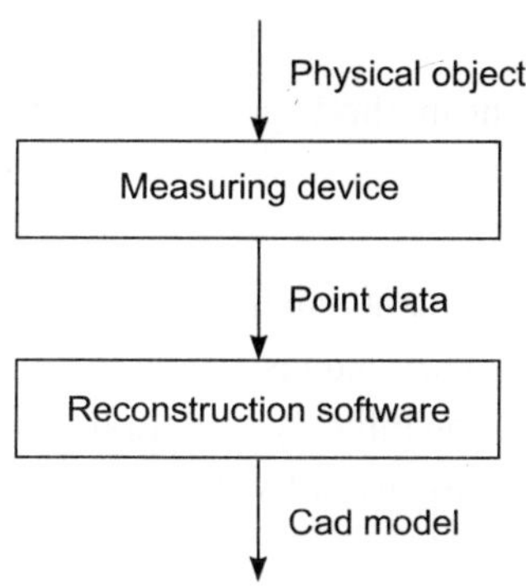


**Figure 5.1** Steps in reverse engineering

Acquiring point data is carried out by an appropriate hardware and the construction of 3D model from the acquired data is carried out using a suitable software.

## 5.2 ACQUIRING POINT DATA

An example of the CoPs generated for a drinking glass is given in Figure 5.2. The methods used to produce CoPs or point data as output can be divided into two major types:

(i) Contact type

(ii) Non-contact type.

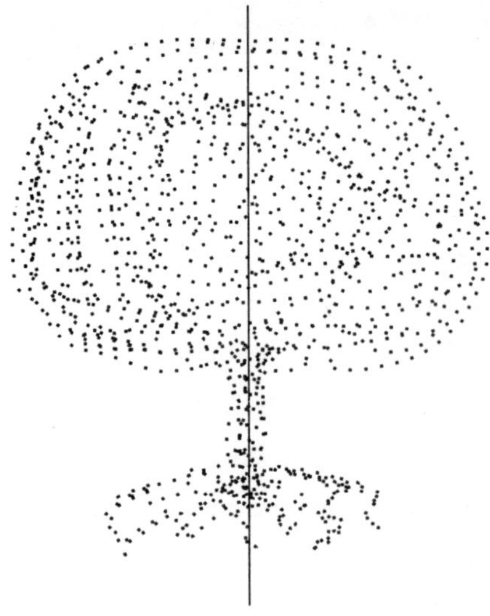

**Figure 5.2** Cloud of points of a drinking glass

Touch probe based systems such as Coordinate Measuring Machines (CMMs) are contact type whereas laser scanners, optical fringing photogrammetry systems come under non-contact type. While the contact types are more accurate, they are slow, labour intensive and they cannot reach interior locations. On the other hand, the non-contact type acquisition systems can produce millions of points within a short time but here the skill of the person doing the scanning is a crucial factor.

### 5.2.1 Contact Type Methods

The various contact type measurement methods are discussed in this section.

#### Manual Measurement

Manual measurement using simple instruments such as scale, measuring tape, vernier calipers, micrometer, bore gauge, height gauge, templates, slip gauges, standard pins etc. will be enough for objects with simple geometry. This is the most labor-intensive method. Furthermore, this method does not create 3D data directly; one needs to use the measured values to create 3D model.

## Touch Probe Measurement

The surface of the object is measured by using contact sensitive sensor or touch probe mounted on Coordinate Measuring Machine (CMM) shown in Figure 5.3. When the probe touches the surface, a signal is sent back to the machine, which records the x, y, and z coordinates of the contact point. The CMM is generally a portal frame type and fixed in a particular location with an environment controlled for temperature, humidity and dust. Depending on the investment, it may have features to automatically scan the object in a predetermined path.

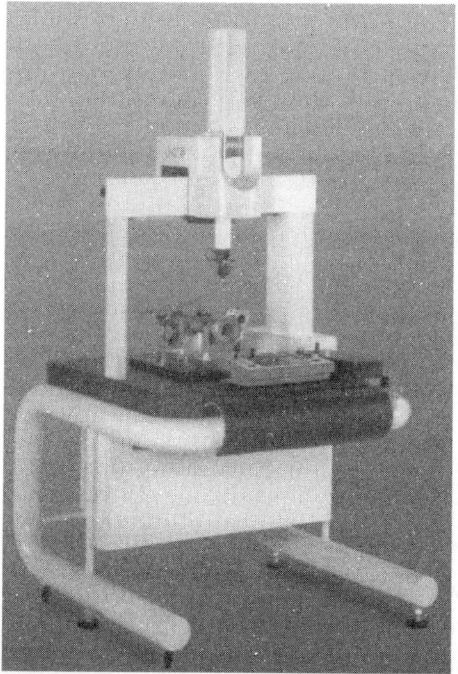

**Figure 5.3** Coordinate measuring machine

The advantage of this method is that the object to be measured does not always have to be moved for full 3D scans since the conventional CMM will have multiple probes calibrated at different orientations and the articulated one is inherently dexterous. Additionally, by scanning only corners and edges, it is easier to recover the topology of the 3D object. Internal openings pose less of a problem. However, this method is labour intensive and time consuming. Furthermore, this method can generate only smaller data sets, typically around 120 points per minute only.

There are also portable CMMs, which look like an articulated mechanical arm. Its articulated arm generally consists of 3 parts, viz., the base, elbow and wrist. The wrist that holds the calibrated probe can be moved manually. Data are collected at the probe tip, most often through the use of a manual switch or button. A portable CMM requires either an integrated or separate controller unit that calculates the probe position based on the angles of the encoders at each joint and the lengths of each link. The x, y and z values are then transferred via a serial line to an application software executing on a computer.

*FAROARM*, a portable CMM developed by FARO, USA, with six or seven degrees of freedom, captures the 3D data of the solid model or existing part when an engineer traces the surface of the part (Figure 5.4). As the stylus of the arm is moved from one spot to another on the object, the engineer presses a button on the handle of the arm to record the position in the form of (x, y, z) and orientation in the form of direction cosines (i, j, k) of the probe. Data can be gathered as single points, streams of points or as complete surface features.

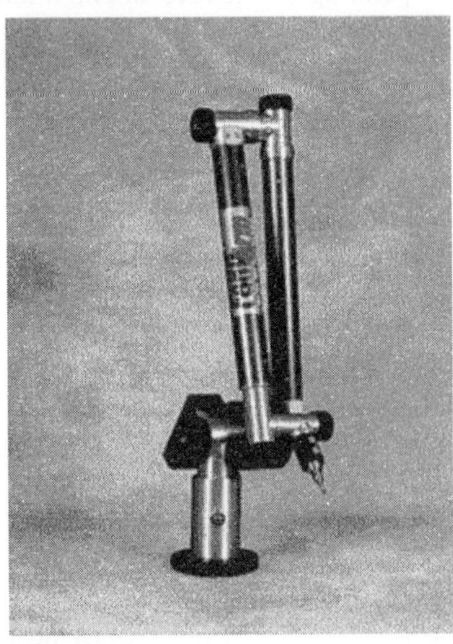

**Figure 5.4** Faro arm — A portable CMM

The arm is paired with software called *AnthroCAM* that gives users the ability to manipulate data, including the capability to create 3D CAD drawings or bring point-to-point data into the system as scanned surface images. Also, the software is compatible with common data formats, including DWG, DXF and IGES. As the engineer scans a part, the image begins to take shape in real time on the computer.

In a typical application, the engineer may take 100 or more measurements. The process, from solid part to finished drawing takes only a fraction of the time using Faroarm that it did when the work performed manually. For instance, a bracket that took 1.5 hr. to reverse engineer manually took only 15 min. with the arm; a reproduction of an antique car part that took 12 hr. manually, took 2 hr. to reverse engineer using the system.

### 5.2.2. Non-Contact Type Methods

Almost all non-contact type measurement methods described except optical fringing method make use of the *method of triangulation* since the laser emitter, the point on the surface and the laser receptor form a triangle (Figure 5.5). In this method a beam of light strikes the surface, and some of the light bounces toward an off-axis sensor.

As shown in the figure, the tilt angle α and triangulation angle θ are related by Schiempflug condition $\tan\alpha = \tan\theta/M$ where M is the on- axis magnification, i.e., image size/ object size.

When designing an optical triangulation system, it is preferable to have small triangulation angle θ. Furthermore, uniform resolution is preferable throughout the surface being measured.

Sweeping the illuminant can scan a surface. Problems involved in the measurement process are:

- Loss of resolution to defocus
- Large variation in the field of view
- Large variation in resolution.

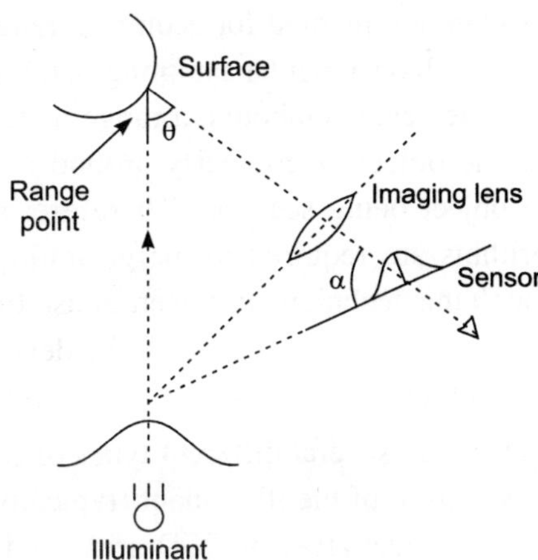


**Fig. 5.5** Triangulation method

Some difficulties are encountered in the triangulation method. These are:

- Finding the center of the imaged pulse is tricky
- Laser width limits accuracy if the surface exhibits variations in the reflectance or shape.

The various non-contact type measurement methods are discussed in this section.

## Laser Scanning

A laser scanner is shown in Figure 5.6. Pulsed laser is directed at the object and the reflections are measured. The time lag between emitting a pulse and receiving it by the detector after reflection from the surface point is the measure of its distance. The resulting CoP has an accuracy of around 0.5 mm. This method produces large data sets (in the order of $10^6$ points) within a very short time. The disadvantages are that the object or the measuring device has to be moved to produce a full 3D scan and interior openings might be in the shadow of the laser beam.

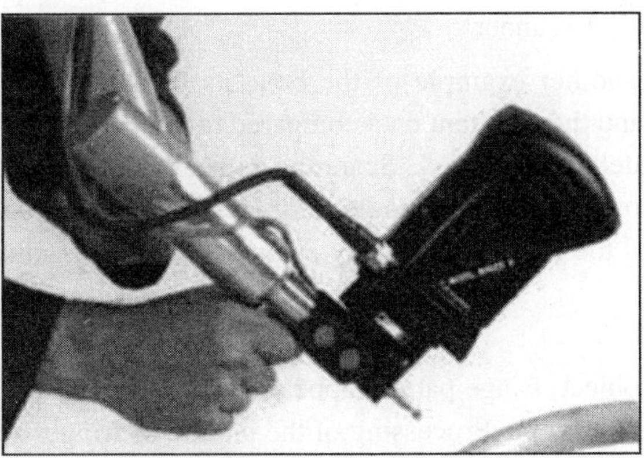

**Figure 5.6** Laser scanner

A common method for acquiring range data is active optical triangulation. Measuring an object's surface depth on a regular sampling lattice produces range data. Then by connecting triangular elements with the nearest neighbours, a range image is created. Generally, a 1D or 2D sensor is swept linearly across the object or circularly around it. This is usually not enough information to reconstruct the entire object being scanned. Therefore, multiple passes must be made from different orientations. Algorithms are required to merge multiple range images into a single description of the surface. Although this technique has been in use for more than 30 years, its speed and accuracy have increased dramatically in recent years with the development of stable imaging sensors such as CCDs and lateral effect photodiodes.

There are several different types of scanners that accomplish this, their primary differences being in the structure of the illuminant (typically point, stripe, multipoint, or multistripe), dimensionality of the sensor (linear array or CCD grid), and the scanning method (move the object or move the scanner hardware). One of the most obvious benefits to laser scanning is the tremendous increase in speed with which a prototype can be reproduced.

Traditional methods call for the object to be measured manually and converted into a CAD model. Not only is this extremely time-consuming, but natural shapes are almost impossible to create using this method. Objects such as an ergonomically designed handle or new toy designs can be easily sculpted and then scanned to ensure the intended result. Laser scanning is at its best when dealing with ergonomic shapes. The entire scanning and post-editing process can happen in hours. This time saving means faster response to requirements. Because laser-scanning technology is relatively quick, it is generally much cheaper than other types of scanning.

Operator experience is a critical factor with optical laser scanning. The operator must follow certain guidelines and be able to predict how the laser will react. Discretion must be shown when viewing the individual scans before merging so that any unacceptable data will be discarded. It is necessary to have a clear understanding of how lasers work so as to know how to deal with them. The lighting in the location, the object's distance from the scanner, and the object's color can all potentially affect the laser scanning process. The operator needs to be able to clearly distinguish acceptable from unacceptable data. The operator needs to be able to read and recognize clearly certain things in the point cloud, when using of laser scanners.

Product verification is another example of the benefits of scanning. After a product has been produced, it can be scanned and the resultant data compared to the CAD/CAM design. The deviation of the part can then be accurately determined. Scanning is routinely used for periodic inspection of multiple parts to analyze how closely the product adheres to the original. This allows for greatly improved quality control and the ability to identify errors in the manufacturing process.

## Optical Fringe Patterns

When light is directed at an object, fringe pattern appears on the surface of the object according to the depth or distance from the light source. Processing of the picture of fringes results in the CoP. The CoP has a higher accuracy than the one generated employing laser technology (0.01mm). The method based on optical fringe patterns also is capable to generate large data sets within short times. The object is usually placed on a turntable to produce full 3D scans. The disadvantage is that it cannot handle interior openings.

### Computer Tomography Scan

Human bodies can be scanned using *Computer Tomography (CT)*. The result is a picture of the human body part in slices. When carrying out image processing of these slices, rings of points can be obtained by marking the interfaces between the various body parts. These rings of points are then combined to form CoPs.CT uses radiation in the form of a highly collimated X-ray fan beam to slice the object. This results in points arranged in parallel planar loops. Standard CT scanners achieve a resolution of 512 × 512 elements within a layer.

### Magnetic Resonance Imaging

*Magnetic Resonance Imaging (MRI)* is yet another technique used in the medical field. MRI images are obtained based on different tissue characteristics by varying the number and sequence of pulsed radio frequency fields in order to take advantage of magnetic relaxation properties of the tissues. MRI differs from CT in at least two key aspects:

(i) MRI measures the density of a specific nucleus

(ii) The MRI measurement system is volumetric, i.e., scanning of the entire body, within the measurement volume, is done all at one time.

CT and MRI represent the finest resolution capability available in diagnostic systems achieving volumetric resolutions. During the scanning process, the patient is taken through the measurement plane 2-3 mm at a time. The information from each plane can be put together to provide a volumetric image of the structure as well as the size and location of anatomical structures. The scanned model becomes a virtual volume that resides in a computer, representing the real volumes of the patient's bone(s).

### 5.2.3 Planing and Partitioning

A thorough understanding of surface creation and the final modelling package is vital in order to make use of the point cloud manipulation software most efficiently. Several scanning hardware manufacturers have developed scanners that accurately digitize amorphous shapes. This allows accurate capture of ergonomic design that interfaces with humans, such as helmets, orthopedic braces, and prosthetic devices. This technology is a fast and safe method for collecting surface information of the human body. Often, manufacturing software can be created for CNC milling right from scan data or from an STL file without taking the extra step of producing a surface model. This means that a prototype can be made and approved, scanned, and a mould made of any proportion quickly and easily. The entire process takes only a matter of days. Scan data can be translated to nearly any file format such as DXF, 3D StudioMax, IGES, STL and Inventor.

Because some objects cannot be scanned in a single pass, it is important to intelligently partition the object into small, discrete sections. After scanning, the discrete sections can be combined electronically into a model representative of the whole project. Surfacing tools are not able to handle overly complex scan data, so it is important to divide the entire object into simple patches that overlap and can be easily scanned and surfaced.

The contoured surface in the Figure 5.7 is partitioned into seven patches. Notice the simplicity of each patch. Each patch benefits by scanning technique unique to its geometry. The keys to dividing up the object are:

***Separate Discontinuities in the Surface*:** Notice that regions 1 and 7 are separated by region 6. This is because region 6 would be considered a discontinuity if regions 1, 6, and 7 were combined.

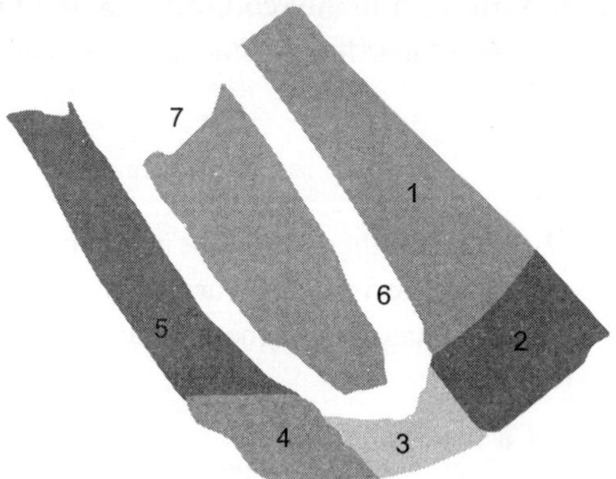


**Figure 5.7** Partitioning and planning

***Separate Discontinuities in the Surface Edges*:** Notice the unique definition of regions 1, 2, 3, 4, and 5. They are separated because, if any two of them were combined, there would be a discontinuity in their edges (a corner) or the trim line of the resulting surface. This is because even powerful surface modelers have trouble with edge discrimination.

***Scan Each Region Meticulously*:** This is one of the most important steps. Often it is helpful to have a picture available and use the "coloring in" method. In the earliest implementation of this technique, this is when an operator watches the graphical interface while scanning and fills in the gaps seen on the screen.

In summary, over scanning and getting too much data is always good. The software can easily filter the data; however, missing data is always bad, and always requires another scan.

## 5.3 CONSTRUCTING 3D MODEL

At the end of measurement, the point data that lie on the boundary of the object only is available. These may be random cloud of points or in some order. If it is available in some order, say as in the case of CT scans, this order must be exploited. One has to obtain the solid model of the object from the point data using an appropriate software. The details of constructing the 3D model are discussed in this section.

### 5.3.1 Procedure

Raw 3D digitized, or point cloud data, is memory hungry, static, and awkward. While it is possible to export raw data directly into CAD software, it can be very painful. Software such as Surfacer, STRIM,

Pro/Scan Tools and Alias are specially developed to handle huge data sets. The following is a description of how such software is used for point cloud manipulation:

***Data Orientation***: Orienting data is the first step. For injection-moulded parts, the data is often oriented relative to the parting plane of the tool. If the part is symmetrical, the point cloud is oriented such that the mirror plane is defined. Once orientation is complete, all data exported to CAD software will be located correctly for easier model creation.

***Manipulation of Data***: Often the final product is a variation of the part that was digitized. Therefore, data manipulation is required to reflect the desired changes. Scaling data is the most common manipulation. Where critical assembly is involved, one may have to obtain the required fits, tolerances and surface finish requirements based on other engineering considerations.

***Extraction of Reference Curves and Base Geometry***: After data orientation and manipulation, information is extracted, one piece at a time, to aid in the CAD model creation.

***Curves***: Cross-sectional and 3D curves can easily be extracted from the point cloud data. The point cloud data can be sectioned through any plane, and a 3D curve can be created through any ridge or feature. In some cases such as CT data, the information is already available in slice form. These planar cross-sectional curves are created quickly. No time is spent smoothing the curves. The curves are only going to be used as templates during the model creation.

***Geometric Features***: Points that make up a flat surface, cylinder, or sphere are isolated, and a best-fit surface is created.

***Verification of Final Surfaces***: Software is used for verification of the CAD model. Surfaces are exported several times during CAD model creation and compared, via a colour variance plot, to the point cloud.

### 5.3.2 Reconstruction of Model

There are several CAD software that convert the available CoP into a 3D object. The data of cloud of points is passed to the CAD software in formats such as IGES, DXF or 3DStudioMax. Various approaches have been used to tackle the problem of solid reconstruction from a given CoP.

Figures 5.8 to 5.14 illustrate an algorithm, which combines the technique of numerical optimization methods with triangulation methods for generating mathematical representations of solids from 3D point data. The solid representation obtained takes the form of an algebraic function whose level surface closely approximates the surface described by the data. The algebraic function is obtained via Implicit Solid Modelling, a constructive scheme for approximating Boolean volume set operations on implicitly defined primitive volumes, and is comprised of a blended union of spherical primitives. The parameters of the algebraic function are the spatial locations and radii of the spheres as well as the parameters that describe the blending of these primitives. Fitting an implicit solid model to a data set is formulated as a sequence of non-linear optimization problems of an increasing number of variables. The cost function we employ in these optimizations is a weighted combination of discrepancies in location (distance from points to boundary of reconstructed object), discrepancies in surface normal, and desired curvature characteristics of the reconstructed solid. Since a set of trivariate data points without any connectivity information is ambiguous, an infinite number of solids, in principle, can be constructed to fit them. Different characteristics of the solid can be specified through the cost function to create the most desirable interpretation of the data. The starting point of the optimization corresponding to the starting configuration of the primitives is determined by performing a triangulation on the data set, and is based on the locations and sizes of the resulting tetrahedron. The effectiveness of the algorithm is demonstrated through the reconstruction of a femur. Trade offs between accuracy and compactness of the representations are also examined. The reconstructed the human femur is shown in Figure 5.14. Figure 5.15 shows a die half obtained by subtracting the femur from a cube. Two such complement die halves form the cavity required to mould the bone implant.

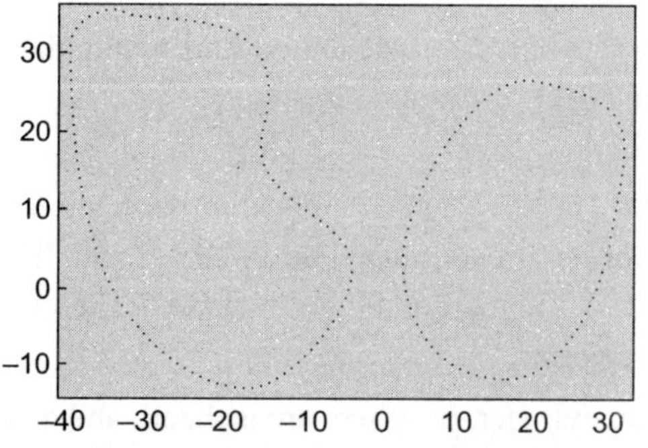


**Figure 5.8** Sliced 3D data

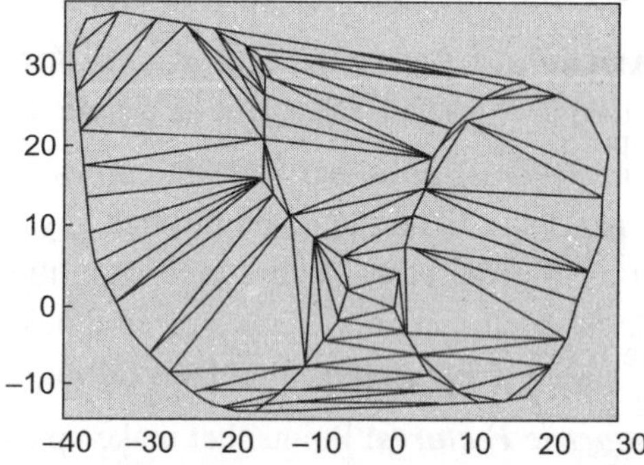


**Figure 5.9** Triangulation

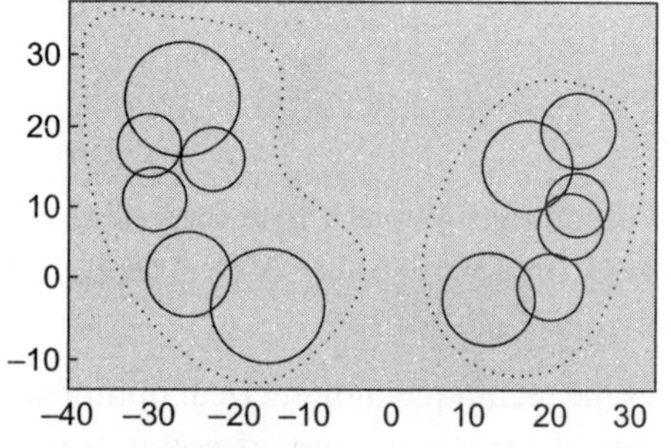


**Figure 5.10** Approximation by circles

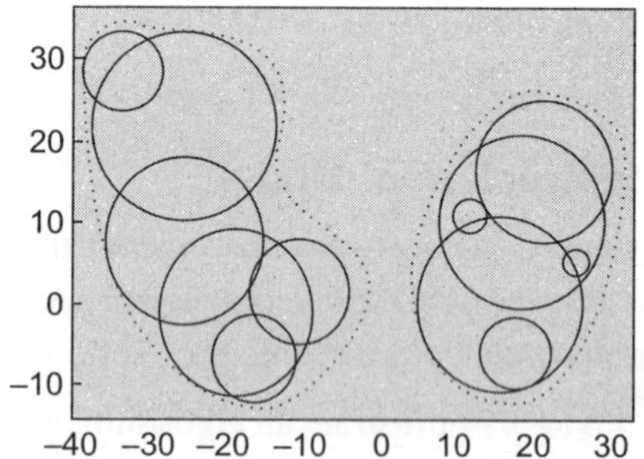


**Figure 5.11** Improved approximation by circles

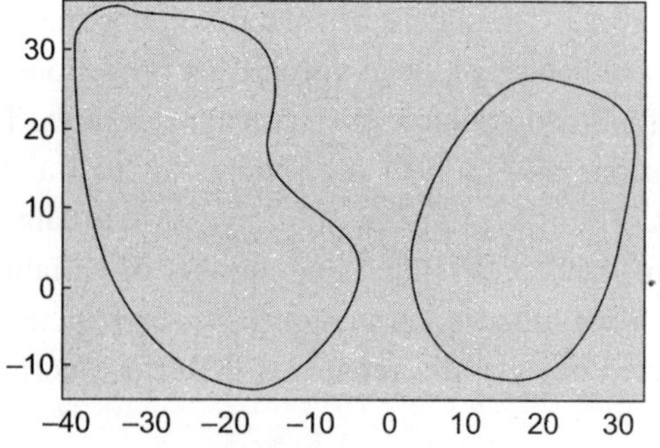


**Figure 5.12** Curve fitting

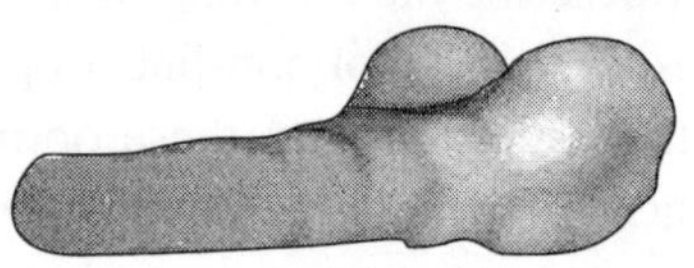

**Figure 5.13** Surface regeneration

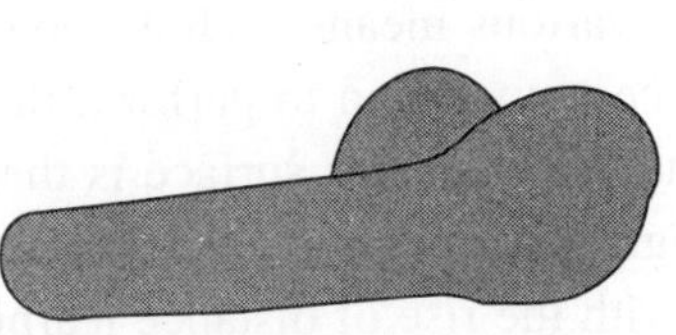

**Figure 5.14** Surface smoothing

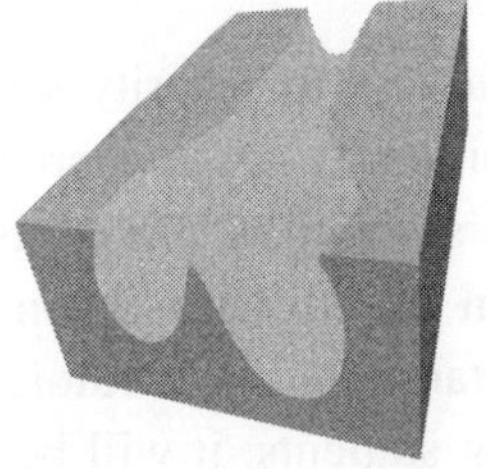

**Figure 5.15** Boolean of Femur and a cube (a die half)

## 5.4 APPLICATIONS

Some important applications of Reverse Engineering are presented in this section.

### 5.4.1 Film Industry

The film industry quite frequently needs models of things or creatures, which do not exist. Therefore, plasticine models are created by hand, which are subsequently scanned. Once the fitted surfaces are available in the computer, the creatures can be brought to life and animated. Examples for this are images created for the films *Godzilla* and *Jurassic Park*. Similar applications can be found across the entertainment industry and the production of commercials. The Figure 5.16 on the left shows the castle after fitting has already been done. The rendered model is shown on the right. This model is now available for cartoons, films or commercials.

**Figure 5.16** Reconstruction of a castle

### 5.4.2 Security

It is common to restrict access to security sensitive areas by various means such as keys, cards, and fingerprint or eye recognition. A novel approach is to use face recognition to perform the same task. The face of the visitor is scanned, a surface is fitted over the CoP and this surface is then compared with one, which has been stored. Only when the surfaces agree, access is granted. Face recognition might also play an important role in the future of education with the rise of distance learning over the Internet. When examining students, it will be necessary to verify that the correct student is sitting in front of the computer answering the test questions. Another security application is the safeguarding of museum pieces. By scanning the exhibits and keeping an electronic copy, it is easier to detect false from original artifacts.

### 5.4.3 Accident Assessment

After road accidents, the police typically measure out the accident spot and the position of the involved parties relative to each other. When using laser cameras, the police simply take a few snapshots of the accident. These pictures are then fed into a computer, which recreates the accident scene. Additionally, it is possible to reconstruct the accident.

### 5.4.4 Product Design/ Rapid Prototyping

Engineering components sometimes need to be styled. Typical examples include the shapes of cars, kettles or telephones. These parts may be as clay or plasticine models. Rather than recreating them in the computer, it is possible to scan them and generate geometric CAD models of them. This styled part can be manufactured using any of the RP processes early in the design cycle.

### 5.4.5 Quality Control

One part of quality control is to compare the manufactured shape to the required shape. For simple-shaped products, this can be done by means of simple measurement. However, this is infeasible for complex, large parts. Using scanning and surface fitting techniques, it is possible to capture and recreate the manufactured shape and compare it with the required shape.

### 5.4.6 Fashion

The fashion industry has started to make extensive use of these technologies. They perform full body scans of people to generate geometric models. This allows designers to create their new clothes directly on the models in the computer. In this way, the designers are able to create a very good image of the appearance of their clothes.

### 5.4.7 Restoration

When carrying out restoration work, there are usually no drawings of the components available. In such cases, the components can be scanned, a geometric model can be built and the parts can be manufactured using modern CNC machines. Similarly, when engineering drawings are lost the CAD models can be reconstructed from an existing part ( Figure 5.17).

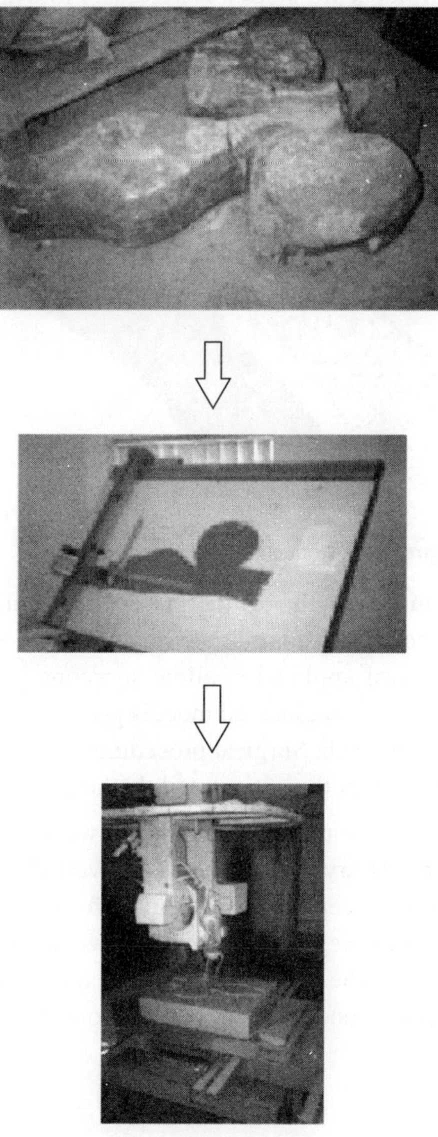

**Figure 5.17** Restoration work

### 5.4.8 Medicine

In medicine, cavities in teeth are scanned, which then enable a filling to be exactly milled from solid ceramics, and glued into the tooth for a longer lasting filling. Another area is the assessment of tumor growth. Sometimes, it can be difficult to judge whether a tumor has grown between two consecutive body scans. When extracting the CoP determining the tumour and fitting a surface to this CoP, it is possible to easily assess tumour growth accurately.

Traditionally, to manufacture orthopaedic shoes it is necessary to wrap the patients foot into plaster, let the plaster harden then remove it. This is not only time consuming but it is also untidy when scanning the foot, it is easily possible to generate a computer-based image of patient's foot. From this, the orthopaedic shoe can be manufactured using CNC machining (Figure 5.18).

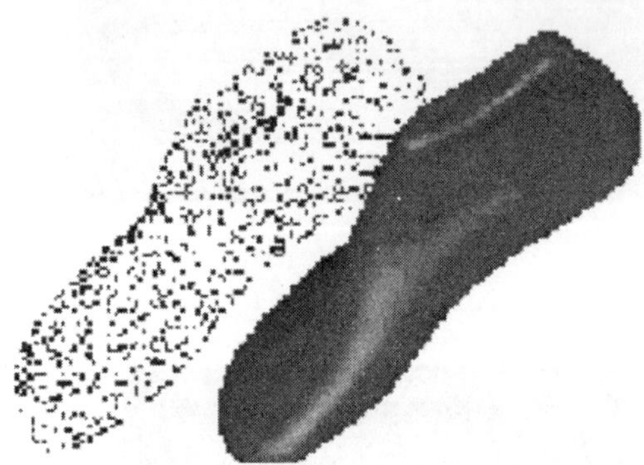

**Figure 5.18** Production of orthopaedic shoes

A precise reverse engineered model facilitates the pre-operative planning of an optimal surgical approach and enables selection of correct or appropriate implants. The reliability and the accuracy of a reverse engineered model in surgical application allow surgeons to rehearse the re-alignment of bones or fitting of implants on the reverse engineered models prior to operating the patient, to evaluate and gain confidence in the planned approach. Surgical procedures continue to be more effective day by day with reduced risk and expense to both the patient and the hospital.

Biomechanical design work is closely related to sculptural work. The human body does not have sharp corners or edges, thus it was necessary to select CAD software that is versatile enough to give the model irregular shape. This software accepts data in neutral formats such as STL and give us the opportunity to interface CTM to CAD. Segmented data can be translated into STL file format and imported into the CAD environment. The CAD environment allows for both the surgeon and the designer to determine critical dimension and mass properties from the CAD model to aid surgeons in their assessment.

### 5.4.9 Simulation

Simulation of parts using numerical methods is only possible when a computational model of the object exist. This is usually not the case for natural objects. These objects can be scanned and a computational object can be created for them. This makes subsequent numerical analyses possible.

## 5.5 CONCLUSIONS

Reverse Engineering in conjunction with Rapid Prototyping & Tooling helps in drastically compressing the product development time. As is clear from the applications and case studies it is useful not only in industries that deal just with machines but it is has found applications in fields as varied as medicine and films.

- The main purpose of reverse engineering is to develop a model that can be used for future reference. It is also used as a technique that helps us to design in a virtual world and therefore saves the process time and cost.
- Realistically, reverse engineering technology can make significant impact in the field of biomedical engineering and surgery. Physical models enable correct identification of bone abnormality; intuitive understanding of the anatomical issues for a surgeon, implant designers and patients as well.

Reverse Engineering is often looked down since someone's design is copied. However, Reverse Engineering and researchers are concerned with only how faithfully they can convert the physical object into a computer model. The ethical and legal responsibilities to refrain from misusing this technology lie with the user. These issues can be ignored considering the benefits it adds to medical, military and other vital applications.

# CHAPTER 6

# Virtual and Augmented Reality

## 6.1 INTRODUCTION

*Virtual Reality (VR)* is a way for humans to visualize, manipulate and interact with the virtual world made up of extremely complex data with the help of computers. The visualization part refers to the computer generated visual, auditory or other sensual outputs to the user of a world within the computer. This world may be a CAD model, a scientific simulation, or a view into a database. The user can interact with the world and directly manipulate objects within the world. Some worlds are animated by other processes, perhaps, physical simulations, or simple animation scripts. Interaction with the virtual world, at least with near real time control of the viewpoint, is a critical test for VR.

The term Virtual Reality can be defined as a computer generated, interactive, three-dimensional environment in which a person is immersed. There are three key points in this definition:

(i) This virtual environment is a computer generated three-dimensional scene, which requires high performance computer graphics to provide an adequate level of realism.

(ii) The virtual world is interactive. A user requires real-time response from the system to be able to interact with it in an effective manner.

(iii) The user is immersed in this virtual environment.

The physical environment around us provides a wealth of information that is difficult to duplicate in a computer. An *Augmented Reality (AR)* system generates a composite view for the user. It is a combination of the real scene viewed by the user and a virtual scene generated by the computer that augments the scene with additional information. The ultimate goal is to create a system such that the user cannot tell the difference between the real world and the virtual augmentation of it. To the user of this ultimate system, it would appear that he is looking at a single real scene.

## 6.2 VIRTUAL REALITY VS. AUGMENTED REALITY

A very visible difference between VR and AR systems is the immersiveness of the system. VR strives for a totally immersive environment. The visual, and in some systems aural hearing sound and other senses are under control of the system. In contrast, an AR system is augmenting the real world scene necessitating that the user maintains a sense of presence in that world. The virtual images are merged with the real view to create the augmented display.

Milgram describes a taxonomy that identifies how AR and VR are related. The real world and a totally virtual environment are at the two ends of this continuum with the middle region called *Mixed Reality (MR)* (Figure 6.1). AR lies near the real world end of the line with the predominant perception being the real world augmented by computer-generated data. *Augmented Virtuality (AV)* is a term created by Milgram to identify systems, which are mostly synthetic with some real world imagery added such as texture mapping video onto virtual objects. This distinction however will fade as the technology improves and the virtual elements in the scene become less distinguishable from the real ones.

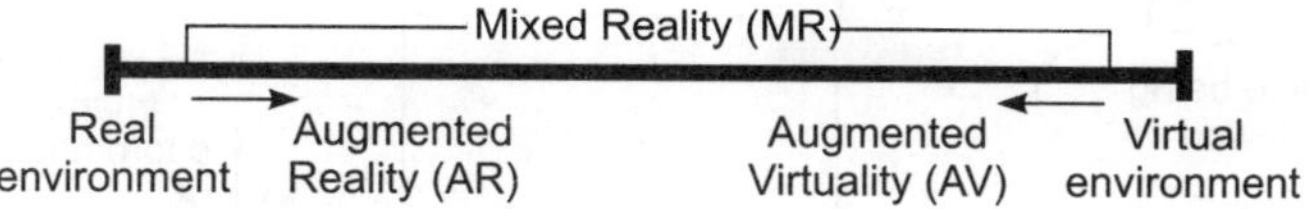


**Figure 6.1** Milgram's Reality—Virtuality continuum

Milgram further defines taxonomy for the Mixed Reality displays. The three axes he suggests for categorizing these systems are:

(i) Reproduction Fidelity

(ii) Extent of Presence Metaphor

(iii) Extent of World Knowledge

Reproduction Fidelity relates to the quality of the computer generated imagery ranging from simple wire-frame approximations to complete photo-realistic renderings. The real-time constraint on AR systems forces them to be toward the low end on the Reproduction Fidelity spectrum. The current graphics hardware capabilities cannot produce real-time photo-realistic renderings of the virtual scene. Milgram also places AR systems on the low end of the Extent of Presence Metaphor. This axis measures the level of immersion of the user within the displayed scene. This categorization is closely related to the display technology used by the system. There are several classes of displays used in AR systems. Each of these gives a different sense of immersion in the display. In an AR system, this can be misleading because with some display technologies part of the "display" is the user's direct view of the real world. Immersion in that display comes from simply having your eyes open. It is contrasted to systems where the merged view is presented to the user on a separate monitor for what is sometimes called a "Window on the World" view. The third, and final, dimension that Milgram uses to categorize MR displays is Extent of World Knowledge. AR does not simply mean the superimposition of a graphic object over a real world scene. This is technically an easy task. One difficulty in AR, as defined here, is the need to maintain accurate registration of the virtual objects with the real world image.

A standard VR system seeks to completely immerse the user in a computer generated environment. This environment is maintained by the system in a frame of reference registered with the computer graphic system that creates the rendering of the virtual world. For this immersion to be effective, the self-centered frame of reference maintained by the user's body and brain must be registered with the virtual world reference. This requires that motions or changes made by the user will result in the appropriate changes in the perceived virtual world. Because the user is looking at a virtual world there is no natural connection between these two reference frames and a connection must be created. An AR system could be considered the ultimate immersive system. The user cannot become more immersed in the real world. The task is to now register the virtual frame of reference with what the user is seeing. This registration is more critical in an AR system because we are more sensitive to visual misalignments than to the type of vision-kinesthetic errors that might result in a standard VR system. Figure 6.2 shows the multiple reference frames that must be related in an AR system.

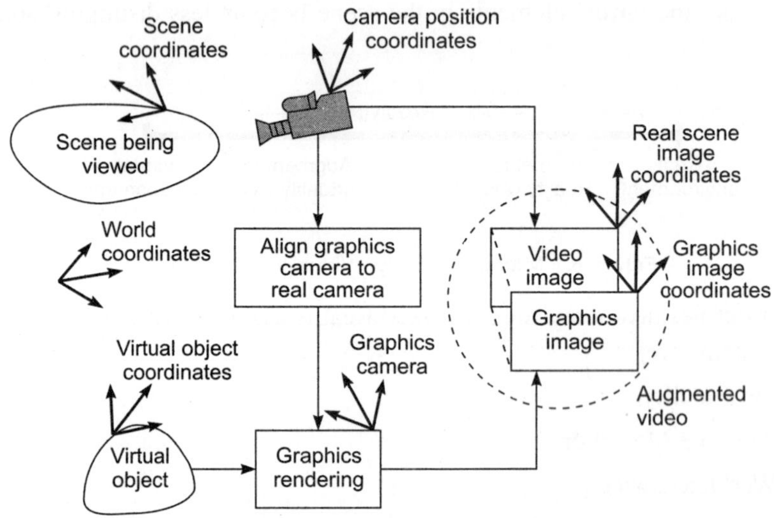


**Figure 6.2** Components of an AR system

The scene is viewed by an imaging device, which in this case is depicted as a video camera. The camera performs a perspective projection of the 3D world onto a 2D image plane. The intrinsic (focal length and lens distortion) and extrinsic (position and pose) parameters of the device determine exactly what is projected onto its image plane. The generation of the virtual image is done with a standard computer graphics system. The virtual objects are modelled in an object reference frame. The graphics system requires information about the imaging of the real scene so that it can correctly render these objects. This data will control the virtual camera that is used to generate the image of the virtual objects. This image is then merged with the image of the real scene to form the augmented reality image.

## 6.3 DEVICES AND TECHNOLOGIES FOR VIRTUAL AND AUGMENTED REALITY

The devices and technologies used for VR and AR are described in this section.

### 6.3.1 Head-mounted Device

Humans in a virtual world must see things as three-dimensional images, otherwise the display will not be convincing. As mentioned, this requires some binocular stereoscopic visual display. The method of achieving this at the moment is by using a bulky headset, rather like a motorcycle helmet. The display must be wide angle, so that people can see things out of the corner of their eye, even when looking straight ahead. The head-mounted device (Figure 6.3) uses a tracking device to detect the head's

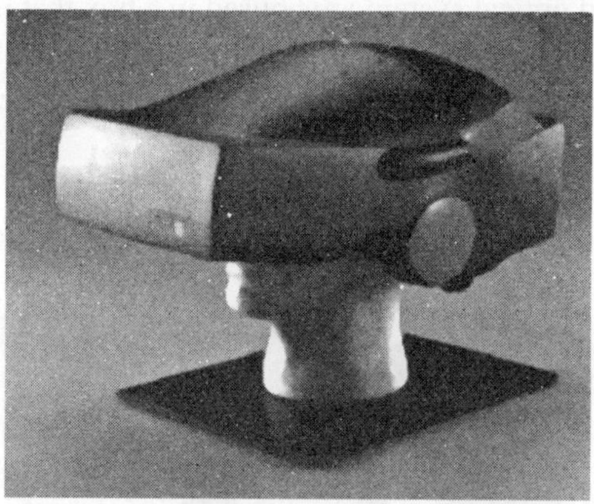

**Figure 6.3** A head-mounted device

movement and will be improved to detect eye movements whilst the person stays still. Such systems are already being experimented with disabled people. The latest generation of headsets is less bulky than the old motorcycle helmets, and looks more like a set of heavy-duty industrial goggles (Figure 6.4), such as someone might use when handling dangerous chemicals. In the future, we can expect VR spectacles, and later on contact lenses, both technologies of which are already in post-R&D phase of development. These systems can provide all the visual clues without the cumbersome weight and represent a major advance. Experimentation is also already underway on projecting images directly into the retina of the eye using very low power lasers.

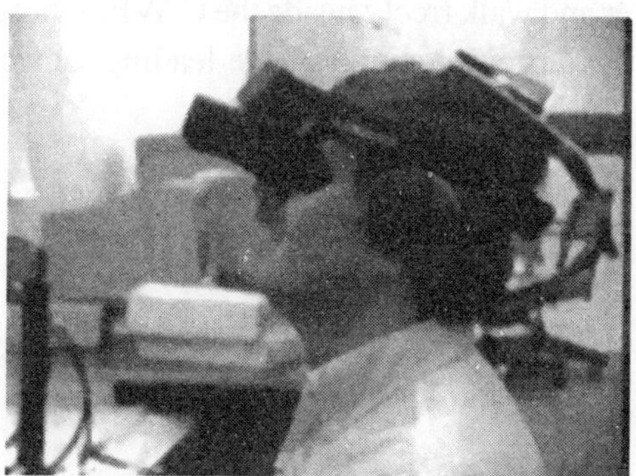

**Figure 6.4** Latest generation headsets

Another fundamental problem with the current generation of headsets is that communication with the VR computer is achieved through direct cable links between the headset and the computer. This limits the mobility of the user. In the near future, wireless communication will overcome this problem. Later on, the computer need not even be nearby. Data from the user as to where he is, and data from the computer as to what he should be seeing, will be delivered using broadband communications networks.

### 6.3.2 Boom

BOOM stands for *Binocular Omni-Orientation Monitor*. It is a head-coupled stereoscopic display device (Figure 6.5). Screens and optical system are housed in a box that is attached to a multi-link arm. The user looks into the box through two holes, sees the virtual world, and can guide the box to any position within the operational volume of the device. Head tracking is accomplished via sensors in the links of the arm that holds the box.

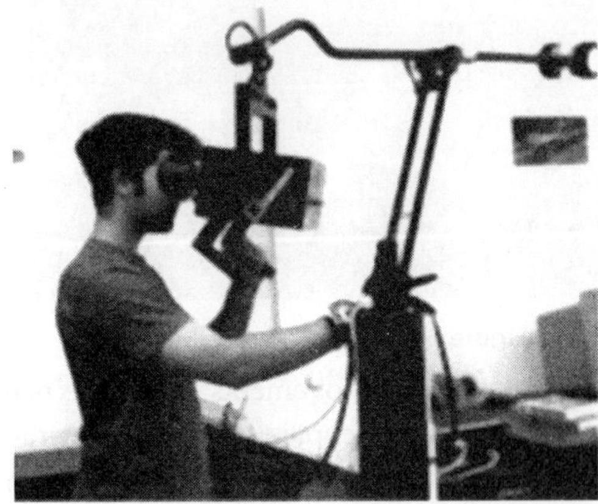

**Figure 6.5** Boom

### 6.3.3 Cave

CAVE stands for *Cave Automatic Virtual Environment*. It provides the illusion of immersion by projecting stereo images on the walls and floor of a room-sized cube (Figure 6.6). Several persons wearing lightweight stereo glasses can enter and walk freely inside the CAVE. A head tracking system continuously adjusts the stereo projection to the current position of the leading viewer.

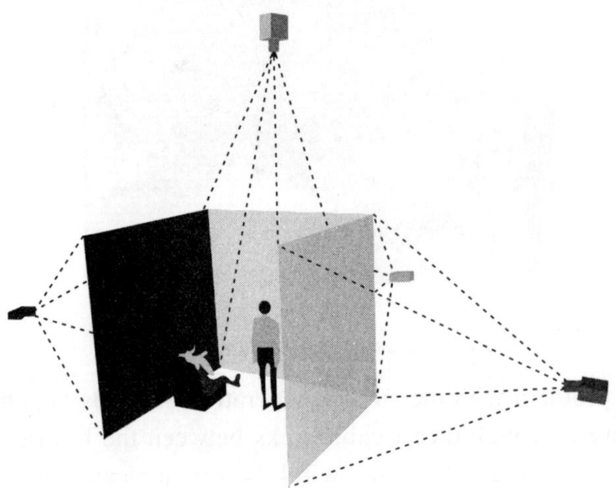

**Figure 6.6** Cave

### 6.3.4 Data Glove

The data glove (Figure 6.7) is worn on the hand, and can then be 'seen' as a floating hand (Figure 6.8) in the virtual space. It can be used to initiate commands. For example, in virtual spaces where gravity does not exist, pointing the glove upwards makes the person appear to fly. Pointing downwards takes him or her safely back to the ground. In this regard, the virtual hand is like a cursor on a standard PC, able to execute commands by pointing at a particular icon and clicking.

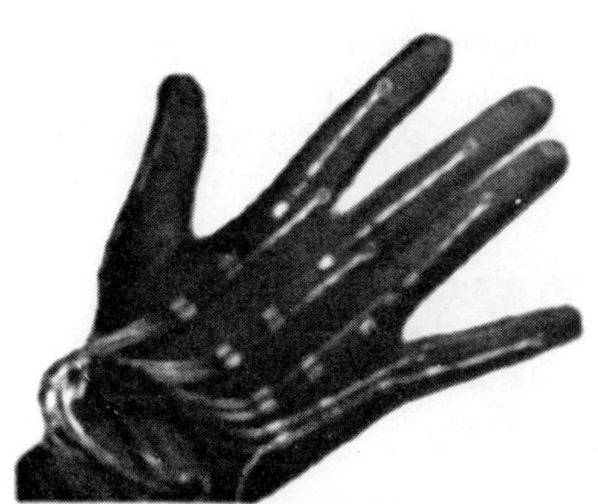

**Figure 6.7** Data Glove

**Figure 6.8** Floating hand moving the steering wheel

The gloves rely on optical fibers to convert the hand movements into signals to the computer. One problem with the glove is the need to provide a sense of touch to increase the haptic (feeling of touch) experience of the wearer. Consider what happens when a human reaches out to grip a virtual object. Although the virtual image shows he or she is gripping an object, the human cannot feel any resistance to the hand tightening movement. Work is under way to achieve this illusion by making the glove resist further closure. The gloves are not without their difficulties. They can be tiring and feel artificial. Indeed, some researchers question the future of the data glove as an input device, though there seems little alternative when it comes to sensual output to the hand.

### 6.3.5 Body Suits

Frequently, users of VR are encased in an entire lightweight body suit. This suit has fiber optic cables or motion sensors at the major joints allowing the VR computers to track the user's movements precisely.

In future, the density of sensors, which can be placed about the body, will increase enabling more accurate portrayals of movement. As with other input devices, enhancements will be made to incorporate output as well. As 'haptic' displays which transmit the sense of touch and force feedback become more prevalent, other physical stimuli may be applicable via such a suit.

### 6.3.6 Sound Generation

Virtual worlds are not necessarily silent, so the person can be surrounded by 3-D sounds, making the experience of the virtual world all the more convincing. At present, sound systems are relatively crude, but much research is underway to create convincing and realistic three dimensional sounds at exactly the right moment, for instance when the data glove hits a VR wall.

VR describes computer generated and three dimensionally depicted spaces with the possibility of real time navigation. A user chooses his or her own perspective view on the shown objects. The visual representation will continuously be adapted to the current movement and field of vision.

### 6.3.7 Display Technologies in Augmented Reality

The combination of real and virtual images into a single image presents new technical challenges for designers of AR systems. How to merge these two images is a basic decision the designer must make. At one end of the spectrum is monitor based viewing of the augmented scene. This has sometimes been referred to as "Window on the World" or "Fish Tank VR". The user has little feeling of being immersed in the environment created by the display. This technology, shown in Figure 6.9, is the simplest available.

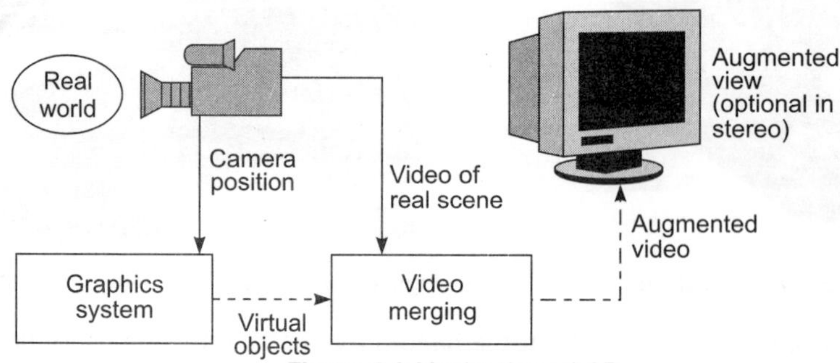


**Figure 6.9** Monitor based AR

To increase the sense of presence, other display technologies are needed. Head-mounted displays (HMD) have been widely used in VR systems. AR researchers have been working with two types of HMDs. These are called *Video See-through* and *Optical See-through*. The "see-through" designation comes from the need for the user to be able to see the real world view that is immediately in front of him even when wearing the HMD. The standard HMD used in VR gives the user complete visual isolation from the surrounding environment. Since the display is visually isolating, the system must use video cameras that are aligned with the display to obtain the view of the real world. A diagram of a video see-through system is shown in Figure 6.10. This can be seen to actually be the same architecture as the monitor based display described above except that now the user has a heightened sense of immersion in the display.

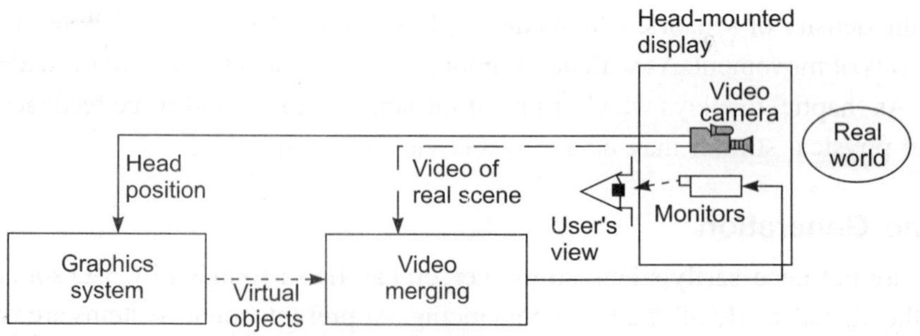


**Figure 6.10** Video see-through AR display

The optical see-through HMD eliminates the video channel that is looking at the real scene. Instead, as shown in Figure 6.11, the merging of real world and virtual augmentation is done optically in front of the user. This technology is similar to heads up displays (HUD) that commonly appear in military airplane cockpits and recently some experimental automobiles. In this case, the optical merging of the two images is done on the head mounted display, rather than the cockpit window or auto windshield, prompting the nickname of HUD on a head.

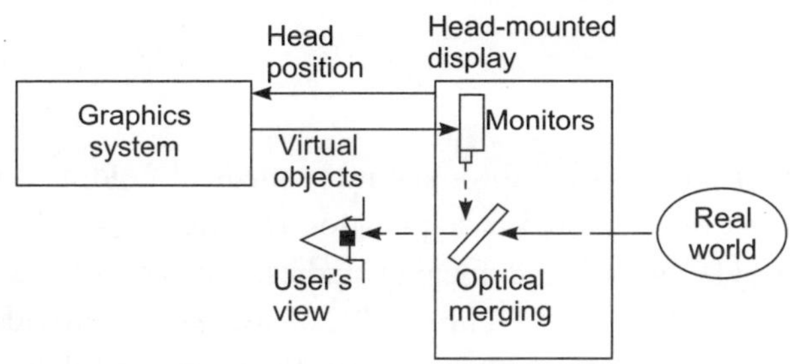


**Figure 6.11** Optical see-through AR display

There are advantages and disadvantages to each of these types of displays. There are some performance issues, however, that will be highlighted here. As both of these displays use a video camera to view the real world, there is a forced delay of up to one frame time to perform the video merging operation. At standard frame rates, there will be potentially a 33.33 ms delay in the view seen by the user. Since everything the user sees is under system control, compensation for this delay could be made by correctly timing the other paths in the system. Alternatively, if other paths are slower, then the video of the real scene could be delayed. With an optical see-through display the view of the real world is instantaneous and hence it is not possible to compensate for system delays in other areas. On the other hand, with monitor based and video see-through displays, a video camera is viewing the real scene. An advantage of this is that the image generated by the camera is available to the system to provide tracking information. The optical see-through display does not have this additional information. The only position information available with that display is what can be provided by position sensors on the head mounted display itself.

### 6.3.8 Modelling Language for Virtual and Augmented Reality

Since 1995 virtual worlds can be accessed via the *World Wide Web (WWW)*. For this purpose, *Virtual Reality Modelling Language (VRML)* has been introduced. VRML allows description of three dimensional objects and sounds. One just needs a browser that interprets the VRML text file and shows it three dimensionally on the computer screen. A VRML world may consist of data stored on various, even far off servers. The graphic interface of the browser allows real-time navigation.

The first version of VRML only included the simple description of objects. Objects can change their position and size in time, e.g. they can be animated. In addition, interactions between object and user are programmable. Using VRML, one can build a sequence of visual images into Web settings with which a user can interact by viewing, moving, rotating, and otherwise interacting with an apparently 3-D scene. For example, one can view a room and use controls to move the room, as he would experience it if he were walking through it in real space.

## 6.4 APPLICATIONS OF AUGMENTED AND VIRTUAL REALITY

Only recently have the capabilities of real-time video image processing, computer graphic systems and new display technologies converged to make possible the display of a virtual graphical image correctly registered with a view of the 3D environment surrounding the user. Researchers working with AR systems have proposed them as solutions in many domains. The areas that have been discussed range from entertainment to military training. Many of the domains, such as medical, are also proposed for traditional VR systems.

### 6.4.1 Medical Applications

Because imaging technology is so pervasive throughout the medical field, it is not surprising that this domain is viewed as one of the more important application for AR. Most of the medical applications deal with image-guided surgery. Pre-operative imaging studies, such as *Computer Tomography (CT)* or *Magnetic Resonance Imaging (MRI)* scans (Figure 6.12), of the patient provide the surgeon with the necessary view of the internal anatomy. From these images the surgery is planned. Visualization of the path through the anatomy to the affected area where, for example, a tumour must be removed is done by first creating a 3D model from the multiple views and slices in the preoperative study. This is most often done mentally though some systems will create 3D volume visualizations from the image study. AR can be applied so that the surgical team can see the CT or MRI data correctly registered on the patient in the operating theater while the procedure is progressing. Being able to accurately register the images at this point will enhance the performance of the surgical team and eliminate the need for the painful and cumbersome stereo-tactic frames that are currently used for registration.

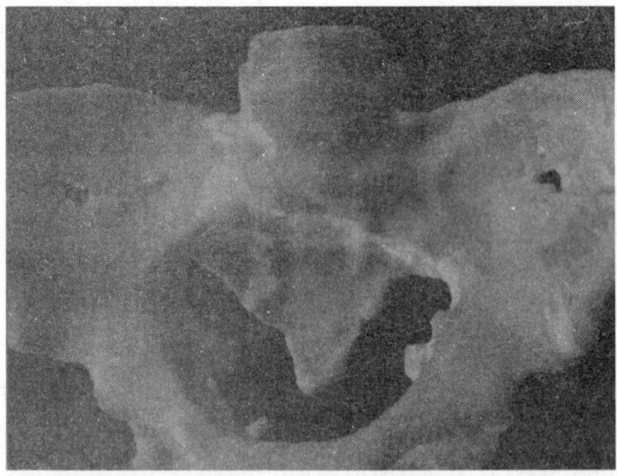

**Figure 6.12** VR image of Pelvis taken from MRI

Another application for AR in the medical domain is in ultrasound imaging. Using an optical see-through display the ultrasound engineer can view a volumetric rendered image of the fetus overlaid on the abdomen of the pregnant woman. The image appears as if it was inside of the abdomen and is correctly rendered as the user moves.

Student doctors can undertake their first operations using VR. The nurses are there, the equipment is there, the anaesthetist is there, the student doctor gives all the instructions to the others and performs the surgery. The operations can be at different levels of difficulty; unexpected complications may be randomly thrown in. There are VR systems that allow surgeons to practice eye and brain surgery before they do the operation on a patient. In addition, VR is used for training in laproscopic (also known as keyhole) surgery.

Using VR, patients requesting plastic surgery are able to assess exactly what the likely changes will be if they undergo a particular type of cosmetic operation. Again, VR allows users to explore 'what if' scenarios until the patient is satisfied, and the cosmetic surgeon then follows the instructions precisely.

Other medical applications include the use of VR in the rehabilitation of patients with brain or spinal cord injuries or with neurological disorders, and for the treatment of patients with phobias.

VR is also used in heart – valve design (Figure 6.13), blood flow visualization, orthopaedic hand modelling (Figure 6.14) and simulation of the human knee (Figure 6.15).

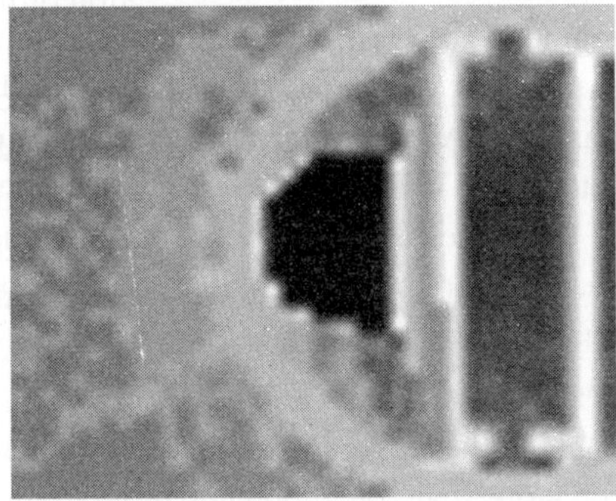

**Figure 6.13** VR Heart valve

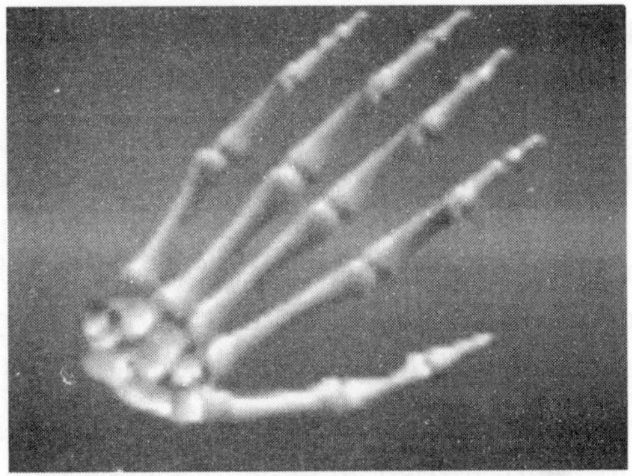

**Figure 6.14** Orthopedic hand modeling

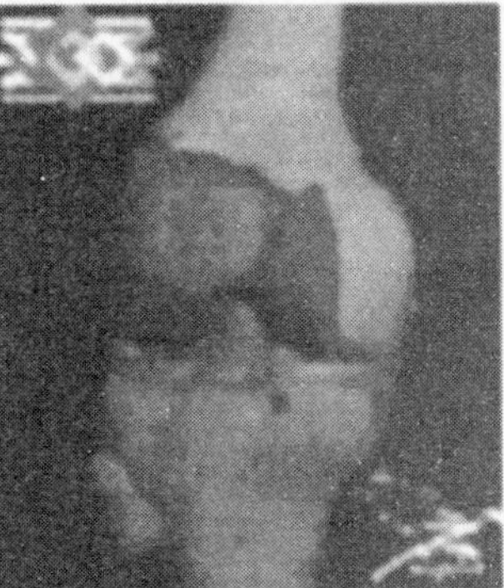

**Figure 6.15** Simulation of the human knee

### 6.4.2 Entertainment and Recreational uses

A simple form of AR has been in use in the entertainment and news business for quite some time. Whenever you are watching the evening weather report the weather reporter is shown standing in front of changing weather maps. In the studio the reporter is actually standing in front of a blue or green screen. This real image is augmented with computer-generated maps using a technique called chroma-keying. It is also possible to create a virtual studio environment so that the actors can appear to be positioned in a studio with computer generated decorating.

Movie special effects make use of digital compositing to create illusions. Strictly speaking with current technology this may not be considered AR because it is not generated in real-time. Most special effects are created off-line, frame-by-frame with a substantial amount of user interaction and computer graphics system rendering. But some work is progressing in computer analysis of the live action images to determine the camera parameters and use this to drive the generation of the virtual graphics objects to be merged.

Obviously there are numerous applications for VR in games. To date, VR games have been confined to arcade machines where the player pays a relatively large amount to experience a relatively short game. Within the next two years, prices will fall sufficiently to allow VR games to be played in the home using multimedia PCs and inexpensive head mounted displays. At the moment, it is this area of games that is pushing VR forward. This has caused concern to many VR researchers, who feel their work is being trivialized, and that the commercial pressures of the games manufacturers are propelling research in the wrong direction. The major arcade games manufacturers are spending considerable money developing a range of games for VR. They believe that there is a large population eager for ever-changing and highly realistic imaginary wars with aliens.

From a commercial viewpoint, the risks in this market include the fickle nature of entertainment consumers, entry barriers in terms of capital requirements for large scale development programmes, the establishment of suitable distribution channels and rapid technology changes where standards issues are still far from resolved.

Sport and entertainment could be transformed by VR. One could be seated in the middle of the orchestra, hearing the three dimensional sounds and watching the performers at work, wandering from place to place amongst them without disturbing them at all. One could watch a play, not from the seat, but from back-stage or on-stage. One could watch a football match from on the pitch.

Training, in particular, could be improved. One could learn how to ski, to play the guitar or to dance. It should be possible, using cameras and motion trackers to record a trainee's attempts and watch actual movement of the limbs, allowing trainers to pinpoint possible areas of improvement. In this way one could improve his golf, netball, snooker or any other chosen skills.

### 6.4.3 Military Training

Not surprisingly, the military sector has been at the forefront of VR research for many years. Even defense cuts in many countries have not harmed its prospects much, as (in theory) a virtual military exercise is considerably less expensive than a real one.

VR is used to simulate battle conditions or can be used to train military personnel on how to use a new piece of equipment, such as portable air defense missile systems. This can be used for general training, or in preparation for a real campaign. Police and security forces already use VR to simulate how to break into a building that is under siege.

The military has been using displays in cockpits that present information to the pilot on the windshield of the cockpit or the visor of their flight helmet. This is a form of AR display. By equipping military personnel with helmet mounted visor displays or a special purpose rangefinder the activities of other units participating in the exercise can be imaged. While looking at the horizon, for example, the display-equipped soldier could see a helicopter rising above the tree line. Another participant could be flying this helicopter in simulation. In wartime, the display of the real battlefield scene could be augmented with annotation information or highlighting to emphasize hidden enemy units.

### 6.4.4 Robotics and Telerobotics

In the domain of robotics and telerobotics an augmented display can assist the user of the system. A telerobotic operator uses a visual image (Figure 6.16) of the remote workspace to guide the robot. Comments of the view would still be useful just as it is when the scene is in front of the operator. There is an added potential benefit. Since often the view of the remote scene is two-dimensional, augmentation with wire-frame drawings of structures in the view can facilitate visualization of the remote 3D geometry. If the operator is attempting a motion it could be practiced on a virtual robot (Figure 6.17) that is visualized as an augmentation to the real scene. The operator can decide to proceed with the motion after seeing the results. The robot motion could then be executed directly, which in a telerobotics application would eliminate any oscillations caused by long delays to the remote site.

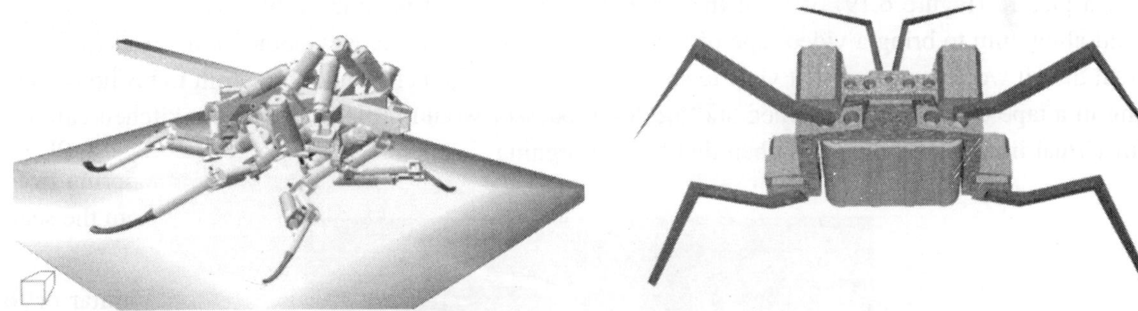

**Figure 6.16** Virtual image of a robot

**Figure 6.17** A virtual robot

### 6.4.5 Manufacturing, Maintenance and Repair

When the maintenance engineer approaches a new or unfamiliar piece of equipment instead of opening several repair manuals they could be put on a display. In this display the image of the equipment would be augmented with comments and information pertinent to the repair. For example, the location of fasteners and attachment hardware that must be removed would be highlighted. Then the inside view of the machine would highlight the boards that need to be replaced. The military has developed a wireless vest worn by personnel that is attached to an optical see-through display. The wireless connection allows the soldier to access repair manuals and images of the equipment. Future versions might register those images on the live scene and provide animation to show the procedures that must be performed. Figure 6.18 shows the wind tunnel experiment being conducted for designing and later manufacturing the object.

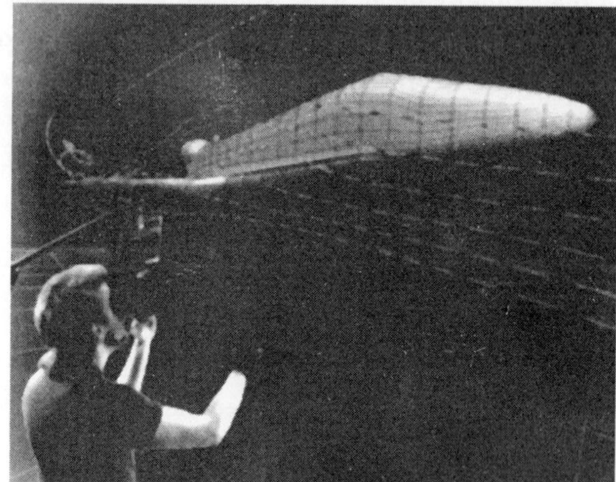

**Figure 6.18** Virtual wind-tunnel experiment

### 6.4.6 Consumer Design

VR systems are already used for consumer design using perhaps more of a graphics system than a VR system. When one goes to the typical home store wanting to add a new floor to his house, they will show a picture (Figure 6.19) of what the floor will look like. It is conceivable that a future system would allow him to bring a video tape of his house shot from various viewpoints in his backyard and in real time it would augment that view to show the new floor in its finished form built to his house. Or bring in a tape of his current kitchen and the AR processor would replace his current kitchen cabinet with virtual images of the new kitchen that he is designing.

**Figure 6.19** A virtual room

Applications in fashion and beauty industry that would benefit from an AR system can also be imagined. If the dress store does not have a particular style of dress in the required size, an appropriate sized dress could be used to augment the image of the customer. As he looks in the three-sided mirror he would see the image of the new dress on his body. Changes in hem length, shoulder styles or other particulars of the design could be viewed on him before he places the order. When he heads into some high-tech beauty shops today, he can see what a new hairstyle would look like on a digitized image of himself. But with an advanced AR system he would be able to see the view as he moves. If the dynamics of hair were included in the description of the virtual object, he would also see the motion of his hair as his head moves.

Architects can use VR to take themselves, or clients, for a walk through the rooms or buildings they are designing. This is one stage better than the images produced by CAD (Computer Aided Design) software. Such applications are achieved by using VR immersion techniques and motion systems capable of making the user move forwards, backwards, to the left and to the right. This allows both the designer and the client to get a real feel for the building (Figure 6.20) that is being proposed. It also allows for 'what if' scenarios - 'what if we were to make that staircase wider?', 'What if we made this room larger?' and so on. The result should be buildings or other facilities that satisfy the client and which are genuinely practical, as they have been experienced by the people who will be using them day to day. The structure is not the only feature that can be tested: alternative materials, colours, lighting (Figure 6.21) and ergonomic factors can also all be tried.

**Figure 6.20** A virtual building

**Figure 6.21** Lighting in a virtual room

### 6.4.7 Science and Engineering

A skilled engineer can put on VR equipment, probably in the form of bifocal spectacles. He can then immediately see how wires are to be connected in an electrical or telephone network or how to repair the car engine. The staff just follow where the wire is to be laid, what things to tighten or to undo, stringing out the real wire, or solder, on the virtual image they can see.

Good visualization tools are a must for CAD/CAM/CAE applications where complicated designs and difficult concepts abound. The use of 3D stereoscopic imaging greatly enhances the level of understanding and visualization of complex wire-frame (Figure 6.22) and solid model parts (Figure 6.23) and designs. Imagine being able to see a part or mechanical design in true 3D depth prior to building expensive prototypes or models. 3D imaging can save time and money during the prototyping phase of a project and can make difficult designs easy to understand during design reviews or other presentations.

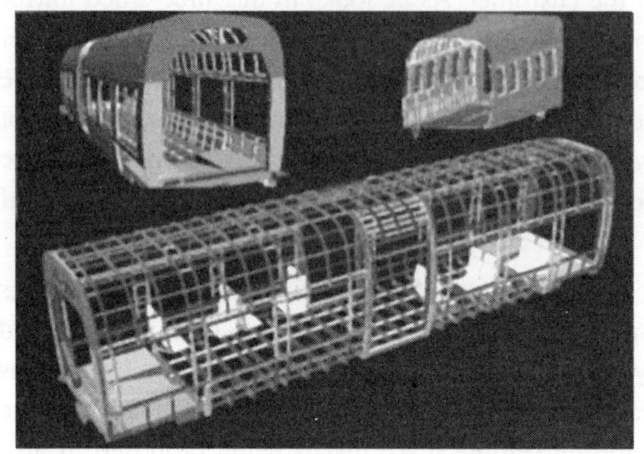

Figure 6.22 Wire-frame model of a railway coach

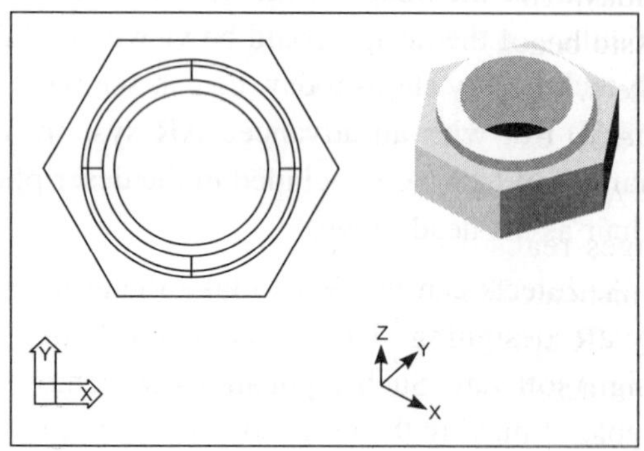


Figure 6.23 Kingpin nut made by CAD

There are other virtual worlds available. One could develop a world in which gravity is half, or twice that of Earth, and explore engineering or other problems. If one wants to know what it is like to travel to the moon and walk on it, VR could provide such an experience. If one wants to know what it is like to be a dwarf, or a giant, VR can provide the necessary sensations. Researchers in relativity have gained new insights into the implications of that theory by using this equipment.

VR is used to simulate a world of molecules (Figure 6.24), to see how they fit into each other and how they react. This is a particularly popular application amongst pharmaceutical companies, keen to explore ideas of which chemicals will bond into which receptor sites. The chemist enters a VR world full of giant molecules, and manipulates one molecule to see how well it will fit into, or dock into, another. This approach is much more intuitive for the chemist, as it converts a three dimensional problem hitherto handled by a two dimensional screen into three dimensions.

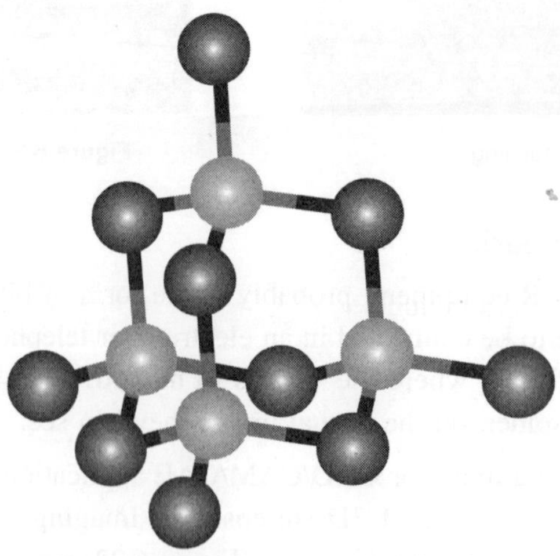

Figure 6.24 Virtual modelling of a molecule

### 6.4.8 Tourism

VR can give clients a taste of what a particular resort is like, the weather, the hotel, the beach and so on. A few people may, in the end, prefer to do their travelling in a virtual world, and no longer travel on holiday. They will order a virtual holiday, which might be to a theme park, an exotic country or a classic beach holiday. Indeed, the tourist trade need not be so conservative as to stick to established centers. VR allows people to go to places that are inaccessible, such as Mars, or indeed to places that do not exist except in the imagination.

A problem for the developers of such systems is the need for realistic images. The creation of images realistic enough to satisfy VR tourists will require a large and detailed database and very sophisticated computer systems. The cost for individuals may remain prohibitive for some time yet, but 'VR holiday centers' could be established using existing graphic workstation technology, spurring on an active new market for creating the visual VR world backdrops and interactive experiences.

### 6.4.9 Education

Virtually every level of education, from school to university, and virtually every subject taught, has potential for VR simulation. Dangerous experiments can safely be conducted in chemistry class, field trips to South American rain forests can be taken without moving an inch, the history lesson can take one back to the French Revolution, instead of just reading Hamlet, one can be transported onto the stage set. VR is already used for teaching calculus and there is a VR version of President Kennedy's assassination so that students can view the event from all angles. Some VR proponents predict that, ultimately, there might be no further need for physical schools or Universities at all. The student could attend a Virtual School or University without leaving home. Experiments already taking place in the USA indicate that this scenario is still some considerable way from being achieved in practice, but important spin-off results apply to distance teaching and life long education. The more general training possibilities include teaching people how to drive cars or to acquire other practical or work-related skills.

### 6.4.10 Applications for the disabled

Someone who is physically disabled could walk, run, jump, do anything an able bodied person could, in the comfort of the virtual world. This could be extended to synthetic speech for the deaf and dumb, and 3D sounds for the blind. Typical applications include allowing disabled people to shop in shopping malls.

Individuals with disabilities comprise one of the most under-employed groups of people. Reasons for their difficulty in obtaining jobs include their limited mobility, reduced manual capabilities and limited access to educational facilities and work settings. VR can help people with disabilities overcome some of these difficulties by facilitating their ability to carry out some of the tasks required in the work setting. VR enhances the accessibility to work by providing the opportunity for individuals to manipulate and explore without requiring physical access, dexterity, and strength. Similarly, those with physical limitations can benefit from the flexibility afforded by an increasingly wide range of VR interfaces, thereby minimizing the limitations of the disability, which can be used as prosthetic mechanisms or as alternative access devices. For example, the head-mounted display can be used as a vision enhancement tool for people with low vision.

Another interesting idea involves the use of 3D "audicons", sounds that surround the participant that can be moved around to control and navigate a computer system. Using a glove interface, users with visual impairments can position the sound in particular places, performing the same "drag and drop" types of functions that are common with visual icons. In one of the most promising developments, biological signals such as the electroencephalogram and the electromyogram have been used to allow persons with quadriplegia to manipulate objects in virtual environments.

### 6.4.11 AR-based Inspection System

Here, we introduce an idea of an AR-based inspection system that augments inspection processes, as shown in Figure 6.25. Conventional VR systems have fatal problems when applying to industrial inspection processes both indoor and outdoor. The major problems should be concerned as follows. When a human operator wears a HMD which is opaque one and displays only virtual scenes.

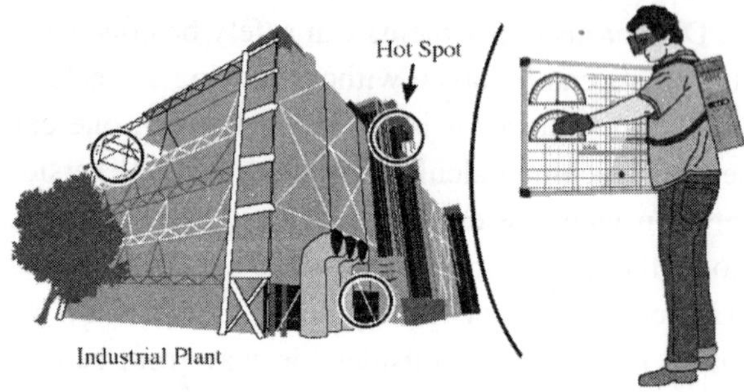


**Figure 6.25** AR based inspection system

- The operator cannot avoid dangerous situations, i.e., closing gates, powerful machine tools or real steps of road, because the HMD is not see-through,
- Usual VR systems cannot be carried out of laboratory, because the measurable area of popular position sensors to be attached to the operator's head is so narrow and limited, such as mechanical arms, ultrasonic sensors, magnetic sensors which need transmitter coil modules and so on,
- Conventional VR systems are too heavy and large to be carried, because they are usually implemented on graphic workstations.

In order to solve the above three problems, the following may be considered:

- In order to walk anywhere and to avoid unexpected dangerous happenings, the operator wearing the HMD needs to see the real environment all the time.
- The position sensors to be attached to the operator's head should be independent and stand-alone.

All equipment to be carried by the operator needs to be implemented in lightweight and compact size.

In order to satisfy the above, the ideal AR-based inspection system should use (a) see-through HMD, (b) inertial navigation and (c) PC-based hardware.

Because the inspection data is provided from a host CAD/CAM station with a large database, the system has to be split into a portable system and a host system with a wireless data-communication channel.

A portable system but not wearable at present includes a complete stand-alone AR hardware. In the system, computer-generated display changes itself depending on the operator's head motion. In other words, the carrying system needs to be able to manage the operator's position and orientation and to generate virtual environments according to them. At present, a tradeoff situation exists between weight and size of graphics hardware, which can be loaded on a carrying system, and the quality of generated virtual environments. Fortunately, the system using the see-through HMD needs to generate not the entire world but only objects of interest.

## 6.5 CONCLUSIONS

There are various applications that could benefit from VR and AR. VR and AR are neither for all users nor for all situations. While VR and AR appear to have significant potential in some cases, it is clear that it must accomplish a particular objective in a substantially better way than traditional techniques. Given the cost of even a moderately complex VR application, the technical support required to operate and maintain it, and, in many cases, the lack of evidence demonstrating its effectiveness, it is fair to ask whether these should be used at all and, if so, under what circumstances and for whom. These are not easy questions to answer. On the one hand, these may save staff time and money and be a superior way to train staff as compared to accomplishing a goal by more traditional approaches. On the other hand, if the skills do not transfer to the real world and if users get bogged down in technical difficulties or experience detrimental side effects, then VR has failed to live up to its promise.

Many users experience physical side effects during and after exposure to virtual environments. Effects noted while using VR include nausea, eye strain and other ocular disturbances, postural instability, headaches and drowsiness. Effects noted up to 12 hours after using VR include disorientation, flashbacks, and disturbances in hand-eye coordination and balance. Many effects appear to be caused by incongruity between information received from different sensory modalities, and to the time lag between the user's movement and the resulting change in the virtual display. These problems are expected to improve with the development of faster workstations and the modification, or elimination of headsets in immersive VR systems.

### 6.5.1 Future of VR & AR

VR & AR have been the subject of considerable research. This industry is immature and there are many social and ethical problems associated with the technology. Bearing in mind the considerable advances in computer technology over the past 30 years, however, it is reasonable to assume that VR & AR will be commonplace in the next 20–30 years.

There will almost certainly be major opportunities for content providers in the information industry in the use of VR, but it may be some time before they emerge clearly. Providing the extensive content for nearly all VR "experiences", both those which mimic nature and those which are pure fantasy, will require tremendous investment in digital production technology, continued expenditure on R&D into areas such as visualization techniques and sound technology, plus training in the use of that technology and in its application to myriad end products. In view of the human capacity to rapidly identify trends and anomalies in visual representations, VR will in the medium term become a key component in the methods for handling the vast masses of electronic data and information that will be available to clients.

With their proven track record in the innovative use of CAD and multimedia, and a conducive legal and regulatory environment, U.S.A. and European companies are well placed to further develop VR. The fact that the VR market is largely language independent is an added bonus. Thus, there is every reason to hope that these companies will play a major part in the worldwide VR market in years to come.

Information industry companies should maintain a close watch on this exciting new technology. In particular, real-time information companies should already be developing VR applications for their clients, not overlooking new developments for using VR over the Internet.

CHAPTER 7

# Computer-aided Engineering

## 7.1 INTRODUCTION

*Computer-Aided Engineering (CAE)* is a technology concerned with the use of computer systems to analyze CAD geometry, allowing the designer to simulate and study how the product will behave so that the design can be refined and optimized. It aims to explain computer technology and its relation to the design process, and to provide an understanding of the functioning and management of a modern design environment.

CAE tools are available for a wide range of analyses. Some of them are listed below:

- Design and Optimization
  - Geometric Modelling
  - Design Sensitivity Analysis
  - Multidisciplinary Design Optimization
  - Optimization Algorithms
  - Reliability-Based Design Optimization
  - Topology Optimization
- Kinematics and Dynamics
  - Vibration Modelling and Simulation
  - Real time Dynamics and Haptics (sense of touch)
  - Mechanisms and Robotics
- Solid Mechanics
  - Biomechanics
  - Composites
  - Computational Mechanics
  - Probabilistic Mechanics and Reliability
- Human-System Interaction
  - Digital Human Simulation and Ergonomics
  - Human Interaction with Advanced Technology and Automation

- Human Computer Interaction and Virtual Reality
- Computational Models of Human Performance

Probably the most widely used method of computer analysis in engineering is the *Finite Element Analysis (FEA).* R. Courant first developed FEA in 1943. He utilized the Ritz method of numerical analysis and minimization of variational calculus to obtain approximate solutions to vibration systems. Shortly thereafter, a paper published in 1956 by M. J. Turner, R. W. Clough, H. C. Martin and L. J. Topp established a broader definition of numerical analysis. The paper centered on the "stiffness and deflection of complex structures". Till the early 70's, FEA was limited to expensive mainframe computers generally owned by aeronautics, automotive, defense, and nuclear industries. The rapid decline in the cost of computers and the phenomenal increase in computing power have made it possible to run FEA software to an incredible precision on PCs as well. Present day super computers are now able to produce accurate results for all kinds of parameters.

## 7.2 APPLICATIONS OF FINITE ELEMENT ANALYSIS IN ENGINEERING

It is not surprising that this method was first used for the stress analysis of aircraft structures because sophisticated stress analysis is traditionally emphasized in the aerospace industry. The potential applications of this method were soon recognized by researchers and engineers involved in other disciplines and industries and the rate at which industry has since adopted this method has been staggering.

A complete list of the applications of this method is not possible, however, a few examples including reference sources are presented here.

- Industrial Applications
    - **Aerospace:** Stress analysis of aircraft structural and engine components; aerodynamic and performance analyses; mechanism and linkage analyses; simulation of aircraft response to various applied loads
    - **Automobile manufacturing:** Stress analysis of vehicle structures; dynamic and impact analyses and simulations
    - **Shipbuilding:** Stress analysis of structures for components and assembled products; hydrodynamic performance analysis
    - **Nuclear power:** Thermomechanical stress analysis of reactor components; thermohydraulic performance analysis of systems; simulation of normal operating and accident conditions
    - **Steel and metal processing:** Stress and thermal analyses of process equipment; prediction of residual stresses and deformation on processed products; macroscopic evaluation of heat treatment procedures
    - **Construction:** Stress analysis of building structures, concrete foundations, underground tunnels, bridges etc.
    - **Resource development and mining:** stress analysis of excavation equipment, geotechnical materials; analysis of response of geological materials to static (e.g., hydraulic) or dynamic (e.g., explosive) loads; thermal mechanical-hydraulic analyses in cases such as geothermal energy development, structural integrity analysis of mine shafts.

- Scientific and Engineering Disciplines
  - Elastic-plastic stress analysis
  - Heat and mass transfer
  - Thermal stress analysis
  - Dynamics and vibrations
  - Fluid mechanics
  - Diffusion and mass transport
  - Soil mechanics
  - Concrete engineering
  - Geomechanics
  - Biomechanics and bioengineering
  - Material science
  - Mechanisms and linkages
  - Physical science

The above list, of course, is by no means complete, and it is only natural for one to assume that it will grow longer and diversify into many more branches of science and engineering applications as time goes by.

## 7.3 THE CONCEPT OF DISCRETIZATION

Many engineering analyses involve physical quantities, such as stress (or strain) everywhere in a solid caused by applied forces (or pressure), temperature at every point in a solid caused by heat sources (or sinks), velocities at every location in a fluid due to difference in heads or pressure, and amplitudes of vibration at different parts of a solid caused by sources of excitation. In FEA, the substances (either solids or fluids) being considered are assumed to be continuous or continua. Therefore, solving problems in continua involves an infinite number of solution points which implies infinite degrees of freedom (DOF). Common sense suggests that the degree of difficulty of the solution to a problem is proportional to the number of DOF involved. FEA that is based on the concept of discretization of continua for converting a problem involving infinite DOF to one with a finite number of DOF is obviously a vital alternative way of solving such problems.

The use of the discretization concept in science and technology is not new. One such common practice was to measure the areas enclosed by closed curves. As illustrated in Figure 7.1 the area of Antarctica may be determined approximately by summing up the areas of the square grids that subdivide the entire domain. The discretization concept is indeed on the principle of integration in calculus, in which the entirety of a physical quantity is determined by summing up all incremental values determined at infinitesimally small increments. The finite difference operators originally formulated by Newton in the $17^{\text{th}}$ century constituted the foundation of a powerful numerical tool, known as the *Finite Difference Analysis (FDA)*, which has been widely used in engineering analyses. An upsurge in the use of this method occurred with the advent of digital computers. FDA, which is built exclusively on the basis of discretization of continua has been used in many engineering disciplines for modern industrial

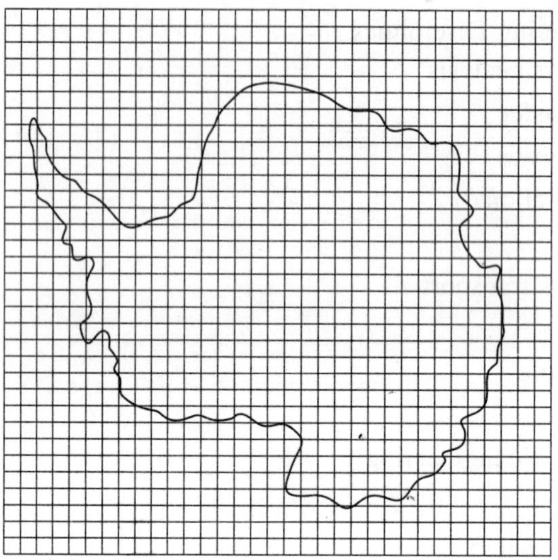

**Figure 7.1** Discretized map of Antartica

applications. Discretization procedures used in FEA are similar to those used in the FDA. Each starts with the subdivision of a continuum into a finite number of sub-domains called elements. These elements are connected at the corners, or in some cases at selected points on edges, called nodes, as shown in Figure 7.2. In this illustration, the solid has a finite shape that can be described by the boundary Ω and is supported at points A and B. The solid is under a set of specified actions (e.g., forces, heat sources) which can be described by

$$\{P\} = P_1, P_2, P_3, ....$$

By virtue of these actions, there exist the required reactions (e.g. deflections, strains, stresses, temperatures) to be described by

$$\{\phi\} = \phi_1, \phi_2, \phi_3, ....$$

Since there are an infinite number of points in the solid bounded by Ω and each point is associated with a set of values $\{\phi\}$, there are obviously an infinite number of $\phi$ values to be determined.

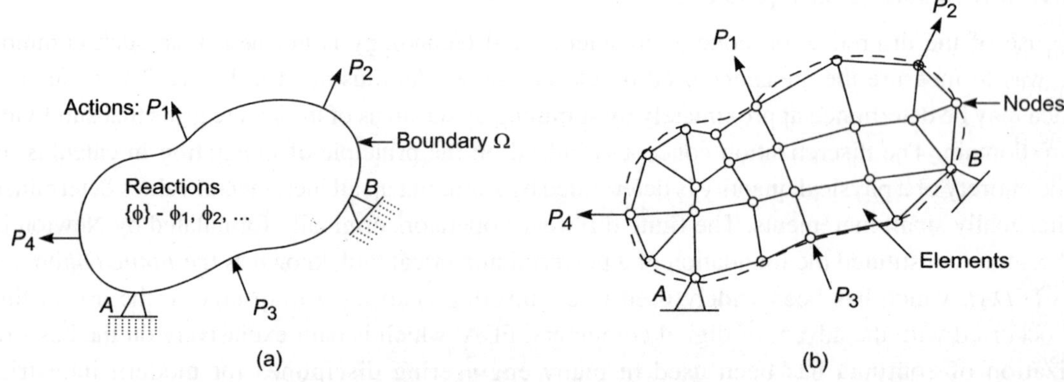


**Figure 7.2** Discretization of a solid (a) Original body, (b) Discretized body

The original solid in Figure 7.2a is subdivided into a finite number of subdomains (elements) with certain shapes (triangular and quadrilateral plates, in this case) and these elements are interconnected at the nodes. After such discretization, the original continuum is no longer a continuous body, but contains a number of discrete pieces connected at the corners and/ or edges in some cases. Another noticeable change is that the originally continuous curved boundary, Ω, has now become segments of straight lines. It is obvious that the discretized body is geometrically similar, but not identical, to the original solid. The degree of similarity of the two bodies, of course, depends on the number of elements used in the discretized body (model). The more elements used in the model, the closer it gets to the original geometry, and thus the result is closer to the exact solution, as illustrated by a simple example in Figure 7.3. Note that the area of a semicircle can be approximated by summing up the areas of the dividing rectangles; one set is entirely enclosed within the circular boundary, and another set is approached from the exterior boundary. Both sets can be shown to asymptotically converge to the exact area as an increasing number of the elements are used in the discretized model.

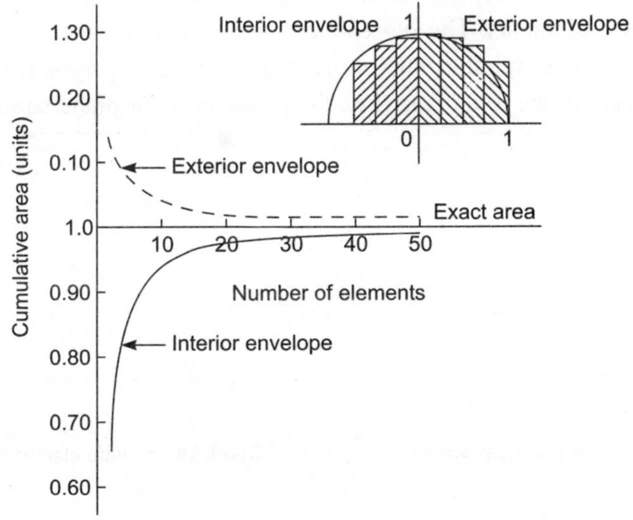


**Figure 7.3** Convergence of an area computation by discretization

The unique advantage of discretization is that after such a process, analysis needs only to be applied to the individual elements of certain simple geometries (e.g. triangular plates), rather than to the entire solid of complex geometry.

The physical quantities, $\{\phi\}$, to be determined in the element can be obtained from these values at the nodes through certain shape or interpolation functions, or

$$\{\phi\}^e = [N]\ \{\phi\}^n$$

where $\{\phi\}^e$ and $\{\phi\}^n$ are the respective physical quantities in the element and the corresponding nodes, and $[N]$ is the shape or interpolation function. The interpolation functions are usually derived on the basis of local coordinates of the element. However, a transformation of this function from the local coordinates to the global coordinate system is performed afterward.

Since the above equation requires solutions only for the individual elements or at the nodes and there are now only a finite number of these elements and nodes in the discretized model, the original problem involving an infinite number of DOF has thus been reduced to a finite number of DOF by means of the discretization process. Solutions of $\{\phi\}$ in the discretized body are hence substantially more attainable.

## 7.4 STEPS IN FINITE ELEMENT ANALYSIS

The wide range of applications of FEA in engineering analysis has been illustrated in the foregoing section. It is unrealistic for anyone to attempt to establish a set of standard procedures for all the computations for the problems described above. However, as a general guideline, most finite element analyses follow the eight steps described below.

### Step 1: Discretize of the Real Structure

Depending on the nature of the structure, a variety of types of elements are available. Figure 7.4 shows typical and commonly used elements. The *bar elements* are frequently used in modelling trusses and frames, and the *plane elements* are suitable when structures have planar geometries or when the variation of the physical quantities is limited to a plane (i.e., the plane stress or plane strain case). For structures

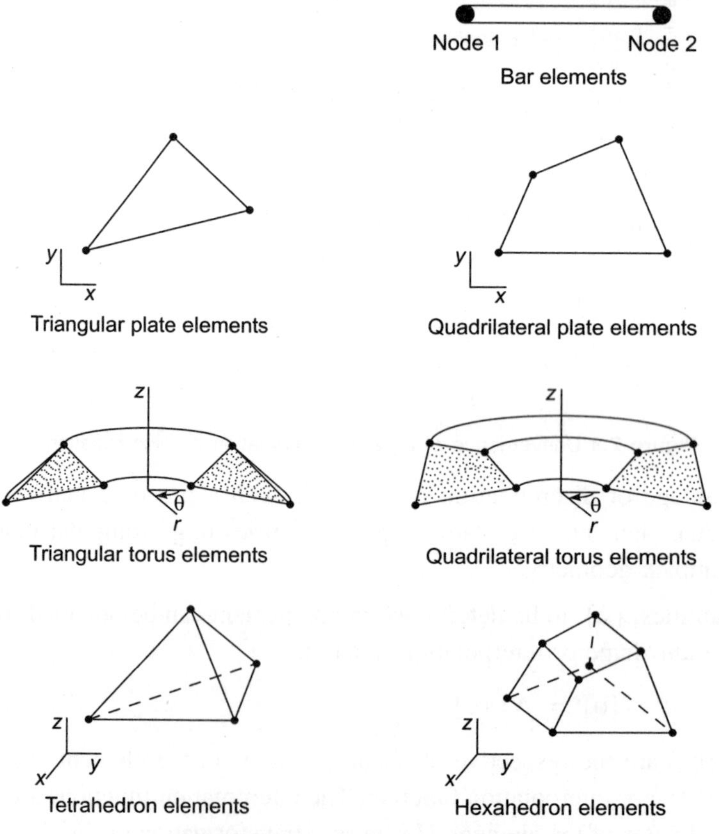


**Figure 7.4** Typical simplex elements

with an axis of geometric symmetry (e.g., cylinders and pressure vessels), the *torus elements* can be used. Finally, for structures with extremely complicated geometry, when a three-dimensional model becomes a necessity, the *tetrahedron* or *hexahedron elements* are commonly used. Many other types of elements (e.g. *beam* and *shell elements*) are developed for special applications. Some elements have additional nodes on the sides which can either be curved or straight. Many isoparametric elements follow this description. Modern finite element computer programmes allow various combinations of different element types, and hence make finite element modelling highly versatile.

## Step 2: Identify the Primary Unknown Quantities and an Appropriate Interpolation Function

Again, depending on the nature of the problem, the primary unknown quantity involved in the solution varies from case to case. One may choose the displacement component as the primary unknown quantity in a stress analysis problem, temperature in a heat conduction analysis, and velocity in a fluid flow problem.

The primary unknown quantity in an element usually is represented by a vector quantity, $\{\phi(\vec{r})\}$, in which $\vec{r}$ represents coordinates [e.g., $(x, y, z)$ in a Cartesian coordinate system or $(r, \theta, z)$ in a cylindrical coordinate system] in the elements. This quantity must be related to the corresponding values at the associated nodes via an interpolation (or shape or trial) function chosen by the analyst. Mathematically, this relationship can be expressed as

$$\{\Phi(\vec{r})\} = [N(\vec{r})]\{\phi\} \quad \text{...(7.1)}$$

in which $[N(\vec{r})]$ is the interpolation function and $\{\phi\}$ are the corresponding nodal values of $\{\Phi(\vec{r})\}$.

Note from the above equation that there is a relationship between the primary unknown quantities in the element and the equivalent values of these quantities at the associate nodes. Conversely, once those unknown quantities at the nodes are solved, the corresponding values at any point within the element can be computed by means of the relationship given in Equation 7.1. There are several forms of interpolation functions that one may use in FEA. One commonly used type is the polynomial function. Polynomials are used because they are easily manipulated in the subsequent computation. The degree of the polynomial chosen is related to the number of nodes in the element, and, to some extent, to the nature of the problems.

## Example 1

Referring to the bar element illustrated in Figure 7.5, two nodes were assigned at $x = x_1$ and $x = x_2$ for node 1 and node 2 respectively. Derive the interpolation function on the assumption that the variation of the longitudinal displacement in the bar element follows a linear polynomial function.

If we let the displacement in the bar element $U(x)$ [equivalent to $\{\Phi(\vec{r})\}$ in Equation 7.1], then the simplest form of polynomial function describing this physical quantity becomes

$$U(x) = \alpha_1 + \alpha_2 x \quad \text{...(7.2)}$$

where $\alpha_1$ and $\alpha_2$ are two arbitrary constants that can be evaluated from the nodal conditions.

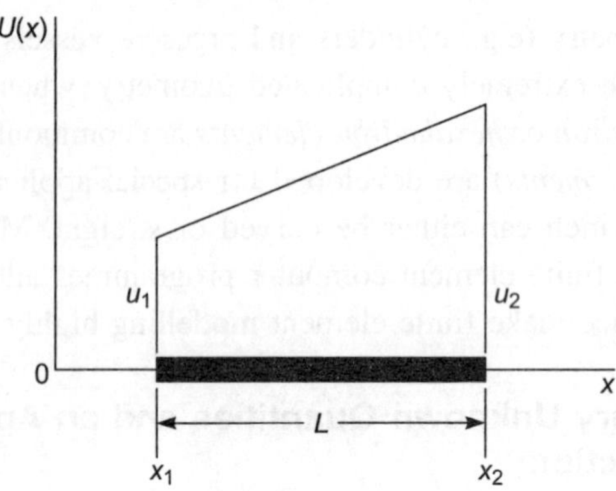


**Figure 7.5** Linear interpolation function for a bar element

We further let $u_1$ and $u_2$ be the respective corresponding displacements at node 1 and node 2. These are unknown quantities at the nodal locations. Note that these are discrete values. The following two simultaneous equations can be derived by substituting $u_1$ and $u_2$ for $U(x)$ in Equation 7.2 with corresponding nodal coordinates $x_1$ and $x_2$:

$$u_1 = \alpha_1 + \alpha_2 x_1$$
$$u_2 = \alpha_1 + \alpha_2 x_2$$

from which one may solve for $\alpha_1$ and $\alpha_2$:

$$\alpha_1 = -\frac{x_2}{x_1 - x_2} u_1 + \frac{x_1}{x_1 - x_2} u_2$$

$$\alpha_2 = \frac{1}{x_1 - x_2} u_1 - \frac{1}{x_1 - x_2} u_2$$

The following relationship between the element quantity, $U(x)$, and the nodal quantities, $u_1$ and $u_2$, can be derived by substituting $\alpha_1$ and $\alpha_2$ in the above expressions into Equation 7.2:

$$U(x) = \left\{ \frac{x - x_2}{x_1 - x_2} \quad -\frac{x - x_1}{x_1 - x_2} \right\} \left\{ \begin{matrix} u_1 \\ u_2 \end{matrix} \right\} \qquad \text{...(7.3)}$$

By comparing Equation 7.3 with Equation 7.1, it can be concluded that the interpolation function has the form

$$[N(x)] = \left\{ \frac{x - x_2}{x_1 - x_2} \quad -\frac{x - x_1}{x_1 - x_2} \right\}$$

## Example 2

Derive the interpolation function for the triangular plate element illustrated in Figure 7.6. The primary unknown quantity $\Phi(x, y)$, in the element can mean a temperature field, pressure variation etc.

Assume the coordinates of the three associate nodes are specified as $(x_1, y_1)$, $(x_2, y_2)$ and $(x_3, y_3)$ with respective nodal values of $\phi_1$, $\phi_2$ and $\phi_3$. Once again, it is assumed that the variation of $\Phi(x, y)$ in the element follows a simple linear polynomial function.

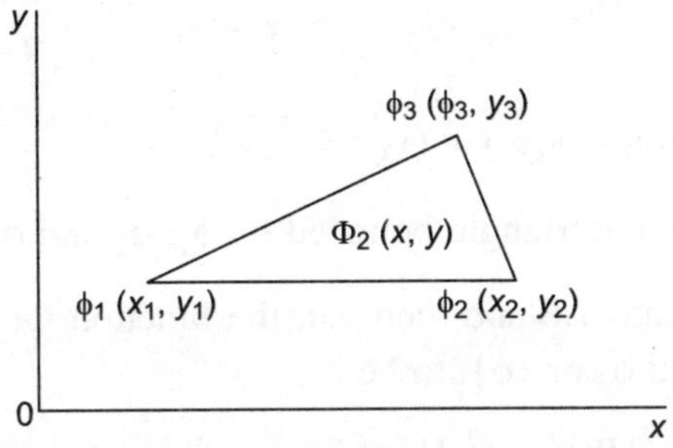


**Figure 7.6** Interpolation function in a 2D domain

Following the element configuration illustrated in Figure 7.6, the function $\Phi(x, y)$ can be assumed to vary within the specified domain according to a linear function:

$$\Phi(x, y) = \alpha_1 + \alpha_2 x + \alpha_3 y = \{1 \quad x \quad y\}\begin{Bmatrix}\alpha_1 \\ \alpha_2 \\ \alpha_3\end{Bmatrix} = \{R(x, y)\}^T \{\alpha\} \qquad \text{...(7.4)}$$

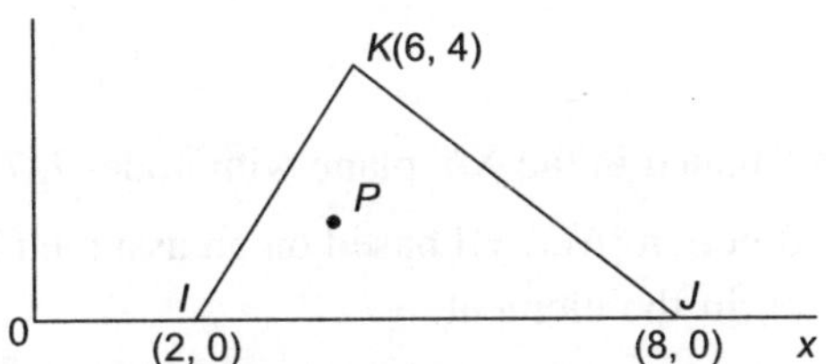


**Figure 7.7** A triangular element

The coefficients $\alpha_1$, $\alpha_2$ and $\alpha_3$ are constants that can be determined by substituting the coordinates of the nodes with specified nodal values $\phi_1$, $\phi_2$ and $\phi_3$ to give

$$\phi_1 = \alpha_1 + \alpha_2 x_1 + \alpha_3 y_1$$
$$\phi_2 = \alpha_1 + \alpha_2 x_2 + \alpha_3 y_2$$
$$\phi_3 = \alpha_1 + \alpha_2 x_3 + \alpha_3 y_3$$

or in a matrix form:

$$\{\phi\} = [A]\,\{\alpha\} \qquad \text{...(7.5)}$$

and

$$\{\alpha\} = [A]^{-1}\,\{\phi\} = [h]\,\{\phi\} \qquad \text{...(7.6)}$$

The matrix [A] in the above expressions contains the specified coordinates of the given nodes as

$$[A] = \begin{bmatrix} 1 & x_1 & y_1 \\ 1 & x_2 & y_2 \\ 1 & x_3 & y_3 \end{bmatrix}$$

The inversion of the above matrix, $[A]^{-1} = [h]$ can be readily performed to give

$$[h] = \frac{1}{|A|}\begin{bmatrix} x_2y_3 - x_3y_2 & x_3y_1 - x_1y_3 & x_1y_2 - x_2y_1 \\ y_2 - y_3 & y_3 - y_1 & y_1 - y_2 \\ x_3 - x_2 & x_1 - x_3 & x_2 - x_1 \end{bmatrix} \quad ...(7.7)$$

where $|A| = (x_1y_2 - x_2y_1) + (x_2y_3 - x_3y_2) + (x_3y_1 - x_1y_3)$

= Twice the area of the triangle bounded by $\phi_1$, $\phi_2$ and $\phi_3$.

By substituting Equation 7.7 into 7.6 and then 7.4, the function $\Phi(x, y)$ can be evaluated by the three nodal quantities by $\phi_1$, $\phi_2$ and $\phi_3$ or $\{\phi\}$, to be

$$\Phi(x, y) = \{R(x, y)\}^T [h]\{\phi\} \quad ...(7.8)$$

The interpolation function for this case can be expressed by comparing the above expressions with that shown in Equation 7.1 and results in the following form:

$$[N(x, y)] = \{R(x, y)\}^T [h] \quad ...(7.9)$$

where the matrix $\{R(x, y)\}^T = \{1 \quad x \quad y\}$ and $[h]$ is given in Equation 7.7.

The interpolation function $[N(x, y)]$ given in Equation 7.9 is a linear function of $x$ and $y$, which is the simplest among all such functions known to exist.

## Example 3

A triangular plane element is situated in the XY plane with nodes $I$, $J$ and $K$ as shown in Figure 7.7.

(a) Derive the interpolation function $[N(x, y)]$ based on an assumed linear variation of the primary unknown quantity, $T(x, y)$, in the element.

(b) If the temperature $T$ has the values, $T_i = 500°C$, $T_j = 400°C$ and $T_k = 300°C$ at the three nodes $I$, $J$ and $K$, what is the temperature at point P in the element with a coordinate of (4, 1)?

We will first evaluate the area of the element to be $A = 12$ units. The interpolation function $[N(x, y)]$ can be expressed by using Equation 7.9 with the $[h]$ matrix to be evaluated following Equation 7.7 as

$$[h] = \frac{1}{2A} = \begin{bmatrix} x_jy_k - x_ky_j & x_ky_i - x_iy_k & x_iy_j - x_jy_i \\ y_j - y_k & y_k - y_i & y_i - y_j \\ x_k - x_j & x_i - x_k & x_j - x_i \end{bmatrix}$$

in which $(x_i, y_i)$, $(x_j, y_j)$ and $(x_k, y_k)$ are the respective coordinates of nodes $I$, $J$, and $K$.

Thus, by substituting appropriate nodal coordinates into the above expression, one may obtain

$$[h] = \frac{1}{12}\begin{bmatrix} 16 & -4 & 0 \\ -2 & 2 & 0 \\ -1 & -2 & 3 \end{bmatrix}$$

The interpolation function for the element can be evaluated to be

$$[N(x, y)] = \{R(x, y)\}^T [h]$$

$$= \{1 \;\; x \;\; y\} \; \frac{1}{12} \begin{bmatrix} 16 & -4 & 0 \\ -2 & 2 & 0 \\ -1 & -2 & 3 \end{bmatrix}$$

$$= \left\{ \left( \frac{4}{3} - \frac{x}{6} - \frac{y}{12} \right) \left( -\frac{1}{3} + \frac{x}{6} - \frac{y}{6} \right) \frac{y}{4} \right\}$$

The temperature distribution in the element, $T(x, y)$, can be expressed as

$$T(x, y) = [N(x, y)] \; \{T\}$$

in which $\{T\}$ represents the temperatures at the nodes. Thus the temperature in the element is

$$T(x, y) = \left\{ \left( \frac{4}{3} - \frac{x}{6} - \frac{y}{12} \right) \left( -\frac{1}{3} + \frac{x}{6} - \frac{y}{6} \right) \right\} \begin{Bmatrix} T_i = 500 \\ T_j = 400 \\ T_k = 300 \end{Bmatrix}$$

$$= 500 \left( \frac{4}{3} - \frac{x}{6} - \frac{y}{12} \right) + 400 \left( -\frac{1}{3} + \frac{x}{6} - \frac{y}{6} \right) + 300 \, \frac{y}{4}$$

The temperature at point $P$ with coordinates of x = 4 and y = 1 thus becomes 433.3°C.

## Step 3: Define the Relationship Between Actions and Reactions

The laws of physics usually provide certain relationships between the *actions* on a substance and the induced *reactions* from the same substance. Table 7.1 indicates some of these physical quantities present in common engineering analysis. The physical law that relates $\{F\}$ and $\{u\}$ in the static stress analysis case is the minimization of potential energy, which is a function of these two quantities. The Fourier law can be used to relate $\{Q\}$ and $\{T\}$ in the heat conduction analysis.

**Table 7.1** Actions and reactions

| Applications | Actions: $\{p\}$ | Reactions: $\{\Phi\}$ |
|---|---|---|
| Stress analysis | Forces $\{F\}$ | Displacements $\{u\}$<br>Strains $\{\varepsilon\}$<br>Stresses $\{\sigma\}$ |
| Heat conduction | Thermal forces $\{Q\}$ | Temperatures $\{T\}$ |
| Fluid flow | Pressure or heads $\{p\}$ | Velocities $\{V\}$ |

### Step 4: Derive the Element Equations

Element equations relate the reaction in terms of the primary unknown and the action that causes the reaction. There are generally two distinct methods used in FEA for the derivation of these equations, viz., *Rayleigh-Ritz method* based on the variational principle and the *Galerkin weighted residual method.* The latter method proves to be more practical for problems that can be completely described by a set of differential equations. Many field problems, such as heat conduction-convention and fluid flow problems, can be handled in this way.

A general form of the element equations can be shown as follows:

$$[K_e]\,\{q_e\} = \{Q\} \qquad ...(7.10)$$

where $[K_e]$ = element stiffness matrix

$\{q_e\}$ = vector of primary unknown quantities at the nodes

$\{Q\}$ = vector of nodal forcing parameters.

### Step 5: Derive Overall Structure Stiffness Equations

This step of the analysis assembles all individual element equations to provide stiffness equations for the entire structure, or mathematically

$$[K_e]\,\{q_e\} = \{Q\} \qquad ...(7.11)$$

in which $[K_e]$ = overall stiffness matrix $= \sum_{e=1}^{m} [K_e]$

$m$ = total number of elements in the discretized model

$\{R\}$ = assemblage of resultant vector of applied nodal forcing parameters

$\{r\}$ = vector of nodal quantities of the entire structure

It should be noted that the entries of the $[K_e]$ matrices common to other elements through nodal connection should be summed up algebraically during the assembly process. This procedure can be demonstrated by the following case illustration.

Referring to the plane structure shown in Figure 7.8, the solid plate is discretized into four triangular elements interconnected at five nodes. If we allow the plate to deform in the *xy* plane only, then each node will be associated with two unknown displacement components along the respective *x* and *y*

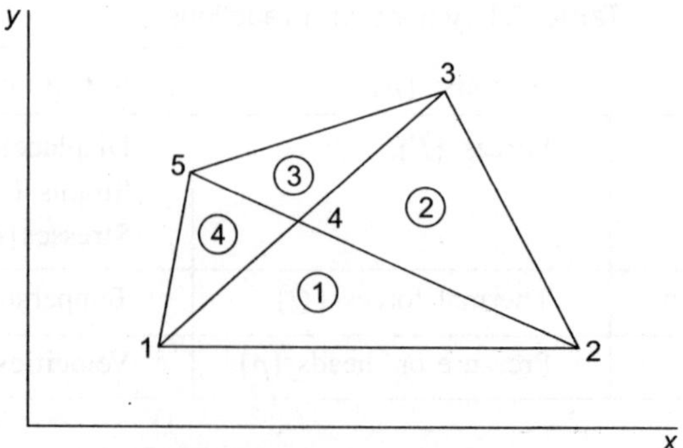


**Figure 7.8** An assembly of element

coordinates. The total number of unknown displacement components in the whole system without any nodal conditions imposed on it is thus equal to 5 × 2 = 10, which is also the total number of degrees of freedom in this case. Since the sizes of the overall stiffness and force matrices are identical to the total number of degrees of freedom involved in the problem, we have thus established the sizes of these matrices for the present case as 10 × 10 for the matrix [*K*] and 10 × 1 for {*R*}.

We shall further recognize the fact that since each node has two degrees of freedom, it is conceivable that the entries under rows 1 and 2 and columns 1 and 2 of the [*K*] matrix should be filled with the corresponding entries of the $[K_e]$ matrix related to node 1. Likewise, the entries under rows 3 and 4 and columns 3 and 4 of the [*K*] matrix correspond to the entries of the $[K_e]$ matrix associated with node no. 2. The same rule applies to the remaining parts of the [*K*] matrix, as well as the {*R*} matrix. The layouts of [*K*] and {*R*} matrices for this particular case are illustrated in Figure 7.9. Closed circles (dots) in this diagram represent the proper positions of the entries of the element stiffness matrices. The overall stiffness and the resultant force matrices are constructed by simple summation of these entries at the same locations.

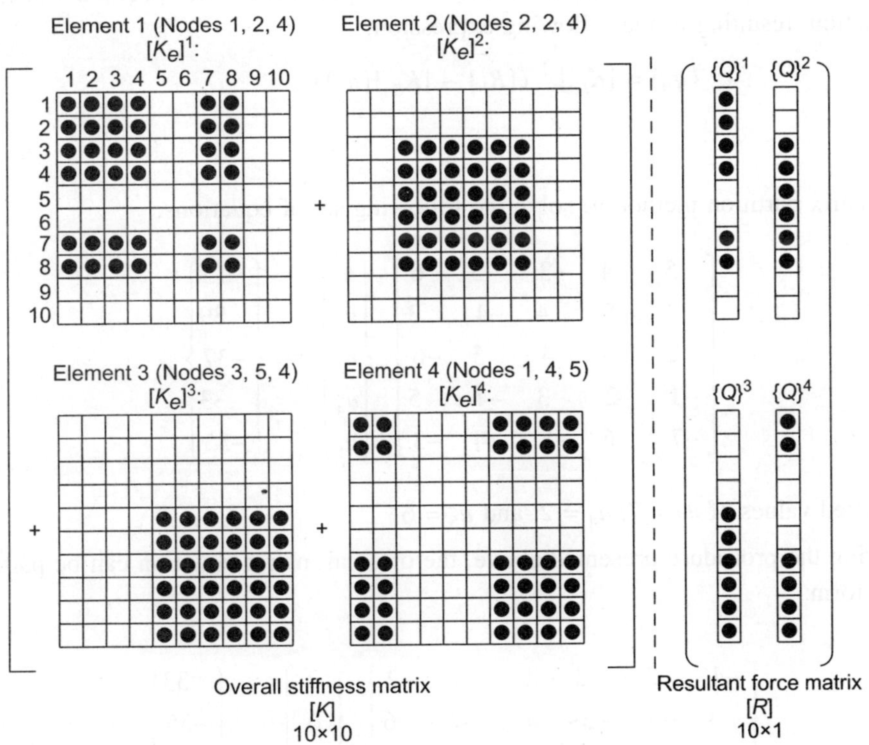


**Figure 7.9** Map of nodal quantities for assembly of element stiffness matrix

The specified boundary conditions of {*q*} or {*R*} are imposed on the assembled equations in Equation 7.11 before solving the equations.

### Step 6: Solve for the Primary Unknowns

It is apparent that Equation 7.11 represents a set of simultaneous algebraic equations. The total number of equations is identical to the total number of primary unknowns, $q_i$ ($I$ = 1, 2,... $n$), at the bodes. Depending on the size and degree of symmetry of the [$K$] matrix, there are generally two methods that can be used to solve for {$q$} [i.e. Gaussian elimination and matrix inversion].

In practical applications, it is desirable to partition the [$K$] matrix and rearrange Equation 7.11 into the following form

$$\begin{bmatrix} K_{aa} & K_{ab} \\ K_{ba} & K_{bb} \end{bmatrix} \begin{Bmatrix} q_a \\ q_b \end{Bmatrix} = \begin{Bmatrix} R_a \\ R_b \end{Bmatrix}$$

where $\{q_a\}$ = specified (known) nodal quantities

$\{R_b\}$ = specific (known) applied nodal resultant forces.

Any interchange of rows of the [$K$] matrix must be accompanied by the interchange of respective columns.

The unknown nodal quantities, $\{q_a\}$, can thus be computed from the specified values by solving the above equation, resulting in the following expression:

$$\{q_b\} = [K_{bb}]^{-1} (\{R_b\} - [K_{ba}]\{q_a\}) \qquad ...(7.12)$$

### Example 4

Use the matrix partition method to solve the following set of equations:

$$\begin{bmatrix} 5 & 4 & -3 & 5 & -6 \\ 4 & -5 & 4 & -4 & 3 \\ 2 & 3 & -4 & 5 & -6 \\ -1 & -2 & 3 & -4 & 5 \\ -7 & 6 & -5 & 4 & -3 \end{bmatrix} \begin{Bmatrix} u_1 \\ u_2 \\ u_3 \\ u_4 \\ u_5 \end{Bmatrix} = \begin{Bmatrix} -15 \\ 39 \\ -37 \\ 33 \\ -55 \end{Bmatrix}$$

with specified values of $u_3 = 7$, $u_4 = 2$, and $u_5 = 6$.

By following the procedure presented above, the original matrix equation can be partitioned into the following form:

$$\left[\begin{array}{ccc|cc} -4 & 5 & 3 & -1 & -2 \\ 4 & -3 & -5 & -7 & 6 \\ 5 & -6 & -4 & 2 & 3 \\ \hline 5 & -6 & -3 & 5 & 4 \\ -4 & 3 & 4 & 4 & -5 \end{array}\right] \left\{\begin{array}{c} 2 \\ 6 \\ 7 \\ \hline u_1 \\ u_2 \end{array}\right\} = \left\{\begin{array}{c} 33 \\ -55 \\ -37 \\ \hline -15 \\ 39 \end{array}\right\}$$

The sub matrices that appeared in Equation 7.12 become

$$[k_{aa}] = \begin{bmatrix} -4 & 5 & 3 \\ 4 & -3 & -5 \\ 5 & -6 & -4 \end{bmatrix} \quad [k_{ab}] = \begin{bmatrix} -1 & -2 \\ -7 & 6 \\ 2 & 3 \end{bmatrix}$$

$$[k_{ba}] = \begin{bmatrix} 5 & -6 & -3 \\ -4 & 3 & 4 \end{bmatrix} \quad [k_{bb}] = \begin{bmatrix} 5 & 4 \\ 4 & -5 \end{bmatrix}$$

$$\{q_a\} = \begin{Bmatrix} 2 \\ 6 \\ 7 \end{Bmatrix} \quad \{q_b\}\begin{Bmatrix} u_1 \\ u_2 \end{Bmatrix} \quad \{R_a\} = \begin{Bmatrix} 33 \\ -55 \\ -37 \end{Bmatrix} \quad \{R_b\}\begin{Bmatrix} -15 \\ 39 \end{Bmatrix}$$

The unknown quantities in $\{q_b\}$ can be solved by using the relationship shown in Equation 7.12 as

$$\{q_b\}\begin{Bmatrix} u_1 \\ u_2 \end{Bmatrix} = \begin{bmatrix} 5 & 4 \\ 4 & -5 \end{bmatrix}^{-1} \left( \begin{Bmatrix} -15 \\ 39 \end{Bmatrix} - \begin{bmatrix} 5 & -6 & -3 \\ -4 & 3 & 4 \end{bmatrix} \begin{Bmatrix} 2 \\ 6 \\ 7 \end{Bmatrix} \right)$$

$$= \begin{bmatrix} 0.12195 & 0.09756 \\ 0.09756 & -0.12195 \end{bmatrix} \begin{Bmatrix} 32 \\ 1 \end{Bmatrix} = \begin{Bmatrix} 3.99996 \\ 2.99997 \end{Bmatrix} = \begin{Bmatrix} 4 \\ 3 \end{Bmatrix}$$

### Step 7: Solve for the Secondary Unknowns

Once the primary unknown, $\{q\}$, is solved, other unknowns may be calculated from the available physical relationships. For example, in the case of elastic stress-deformation analysis, the primary unknowns are the displacement components at the nodes. One may readily envisage that the corresponding element displacement components may be determined by Equation 7.1 with the computed nodal values. The strain components in the element can then be calculated from the displacement-strain relationships following the theory of elasticity. The stress components, of course, can be evaluated by means of the well-known generalized Hooke's law from the computed strain components.

### Step 8: Interpretation of Results

Results obtained from finite element computer codes used to be in tabulated form, indicating the calculated primary and all secondary physical quantities at specified nodes and elements under given applied loads. The analyst could then select critical sections of the body and evaluate these results with respect to the established design criteria as in most engineering design processes. In some instances, the computed results can be expressed graphically. Figure 7.10 shows the contours of isoclinic lines in a gear tooth made of steel. The contours produced by the finite element analysis are shown to give close correlation to those produced by the photoelastic investigation . The critical locations and magnitudes of stress concentration in the gear tooth are readily visible.

**Figure 7.10** Stress contours in a gear tooth by the finite element analysis and photoelastic investigation (a) isoclinics from finite element analysis (b) isoclinics from photoelastic experiment

Expressions of finite element analysis results now take the form of visual displays of the models under different loading conditions. Figure 7.11 illustrates the shape of a cantilever beam under two different levels of loading. Many commercial finite element codes can produce much more sophisticated graphic output, such as animated output as well as comprehensive expression of critical thermal and stress zones in the structure with colour coding.

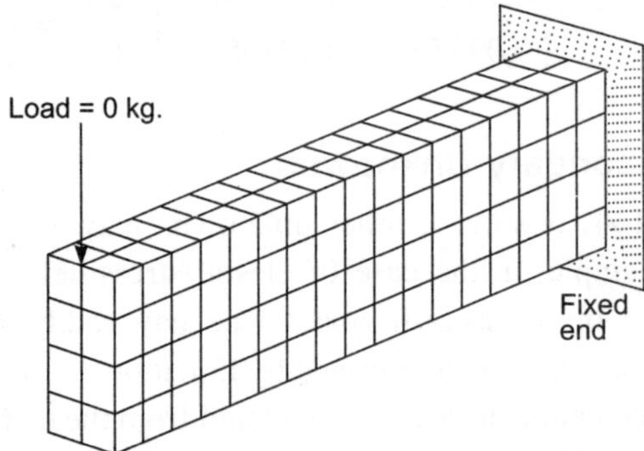


**Figure 7.11** Graphic representation of deflection in a cantilever beam

## 7.5 AUTOMATIC MESH GENERATION

The power and versatility of the finite element method as an effective tool for engineering analyses have been described in above Section. A major drawback of using this method, however, is the laborious and time-consuming task involved in setting up the meshes and entering the information associated with these meshes into the computer. This task becomes even more insurmountable in the cases that involve structures of complex geometries. Automatic mesh generation has become a desirable feature for most commercial finite element programs.

Mesh generation in a finite element analysis involves setting the location of nodes in the model for the structure, numbering the nodes and elements, and specifying the nodal coordinates and nodes that from individual elements. A number of different mesh generators have been developed since the early 1970s.

The primary objective of automatic mesh generation is to minimize manual input to the finite element analysis. It constitutes a major portion of the preprocessor of a finite element program. Computer-generated meshes usually result in fewer errors than what can be accomplished by human effort. Most automatic mesh generators require the user to input mesh topology first. Relevant information such as the topology of the region and the mesh to be generated in the region has to be specified. Desirable mesh shapes in triangular or quadrilateral plane geometries for two-dimensional models, or tetrahedron or hexahedron elements for three-dimensional models need to be entered by the user. Meshes can then be generated based on the user-specified density of nodes, density of element areas, or density of element volumes. The latter scheme is used for three-dimensional analyses. After the mesh generation, the user is expected to make minor local adjustment to rectify any irregularity in the overall mesh. This step may not be necessary if an effective mesh smoothing algorithm is built in the mesh generator.

### 7.5.1 Mesh Generation by Nodal Density Distribution

Many algorithms for mesh generation require the user to specify the density or number of nodes in a specific region of the model. Once the density of nodes is specified, the algorithm can assign proper locations for all individual nodes with sets of coordinates. Elements in the region are then constructed by Boolean connectivity matrices with associated nodes. The numbering of nodes and elements is established by following chronological sequence. The following simple mesh generation scheme is used to illustrate the above procedure.

The region is a plane bounded by vertices *A, B, C* and *D*, as shown in Figure 7.12. A finite element mesh can be generated by an algorithm, with the user specifying the density of nodes along the four

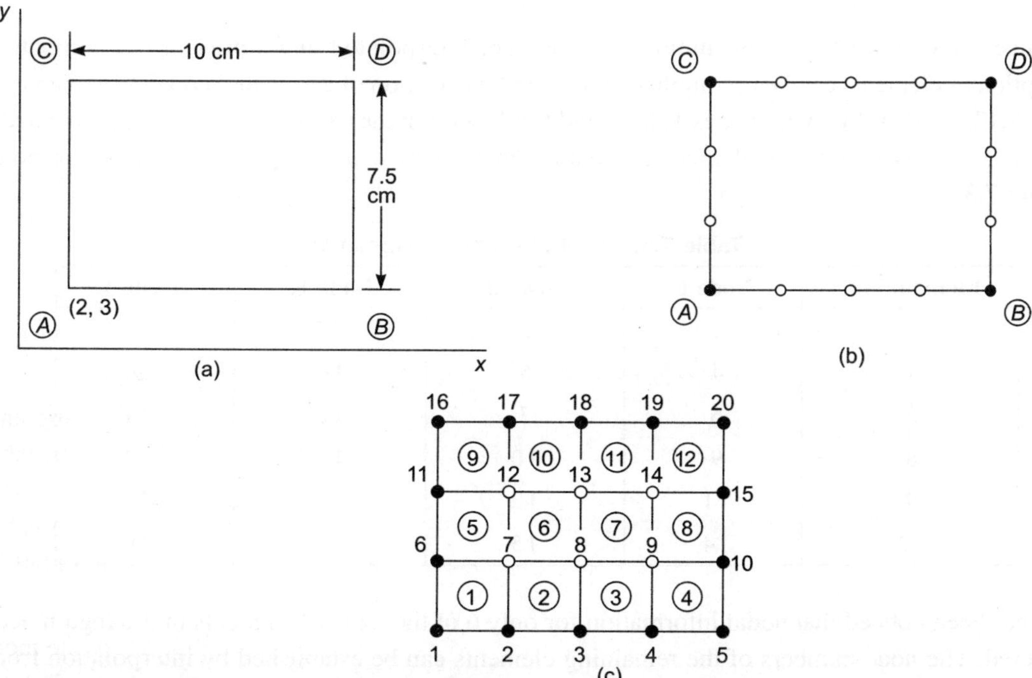


**Figure** 7.12 Mesh generation by nodes: (a) Region to be modelled (b) Generated boundary nodes (c) Generated mesh

edges. Let us assume that five nodes are needed along edges *AB* and *CD*, and four nodes along edges *AC* and *BD*. The algorithm should automatically locate five nodes along each of the edges *AB* and *CD*, with equal spacing. Likewise, four nodes are equally spaced along edges *AC* and *BD*. Since the coordinates of the terminal nodes at vertices *A, B, C,* and *D* are fixed once the coordinate system for the region is set, the intermediate nodes (e.g. nodes 2, 3, and 4 on *AB*, nodes 17, 18 and 19 on CD, nodes 6, and 11 on AC, and nodes 10 and 15 on *BD*) can be readily identified. All interior nodes in the region can be automatically generated by interpolating the coordinates from those of the terminal nodes, as shown in Table 7.2. A typical users input to the program on nodal coordinates is also illustrated in the table.

**Table 7.2:** Input of terminal nodal coordinates

| Node No. | X coordinate | Y coordinate |
|---|---|---|
| 1 | 2 | 3.0 |
| 5 | 12 | 3.0 |
| 6 | 2 | 5.5 |
| 10 | 12 | 5.5 |
| 11 | 2 | 8.0 |
| 15 | 12 | 8.0 |
| 16 | 2 | 10.5 |
| 20 | 12 | 10.5 |

Once the coordinates of all numbered nodes are interpolated and stored in the computer, the description of elements can be accomplished by the identification of associated nodes (e.g, element 1 in Figure 7.12c is identified with nodes 1, 2, 7, and 6; likewise, nodes 8, 9, 14, and 13 are associated with element 7). The user's input of element information in this case can be kept to a minimum, as indicated in Table 7.3.

**Table 7.3:** Input of element information

| Element No. | Node I | Node J | Node K | Node L |
|---|---|---|---|---|
| 1 | 1 | 2 | 7 | 6 |
| 4 | 4 | 5 | 10 | 9 |
| 5 | 6 | 7 | 12 | 11 |
| 8 | 9 | 10 | 15 | 14 |
| 9 | 11 | 12 | 17 | 16 |
| 12 | 14 | 15 | 20 | 19 |

It has been noticed that nodal information for only 6 of the total 12 elements in the region needs to be entered. The node numbers of the remaining elements can be established by interpolation from the numbers of each pair of terminal element numbers, as shown in the first column in the table.

This simple technique of mesh generation can be used effectively in two-dimensional plane geometries. One can expand the use of the above scheme to a more general case of plane geometry, as illustrated in Figure 7.13. In such cases, the interpolation of the coordinates of intermediate and interior nodes requires a slightly more complicated formula that takes into account the slopes of the lines on which all these nodes lie.

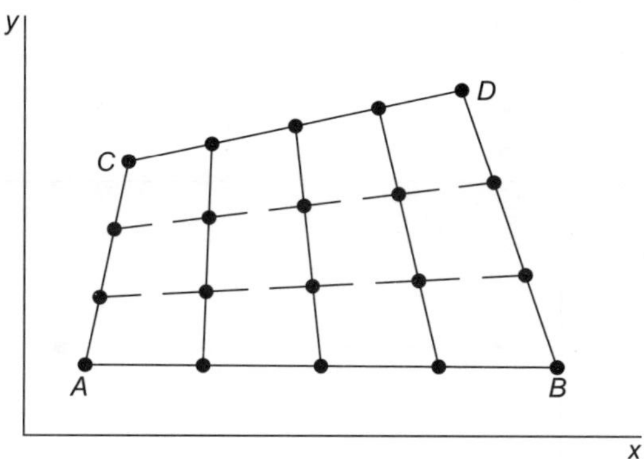


**Figure 7.13** Mesh generation in a quadrilateral region

Many mesh generators produce meshes of triangular shapes, rather than quadrilateral elements as illustrated in the above example. Various methods called triangulation are available for that purpose. A good triangulation algorithm maximizes the sum of the smallest angles of the triangles for optimal analytical results.

### 7.5.2 Mesh Generation by Area Density Distribution

Mesh generation by nodal density distribution works well for two-dimensional analyses involving relatively simple geometries, such as those illustrated in Figures 7.12 and 7.13. A more efficient method is the specification of the density of element areas in the region. The principle of this mesh generation scheme can be demonstrated by a simple example of generating two elements of equal areas in a plane region, as shown in Figure 7.14a.

The reader will find that one way the region ABC can be discretized by two elements with equal areas is depicted in Figure 7.14b. The locations of nodes 2 and 4 can be established by solving for $h_1$ or $h_2$ from the following equations.

$$(b + b_1)h_1/2 = b_1h_2/2 = (bh/2)/2 \qquad ...(7.13)$$

$$h_2/b_1 = b/h \qquad ...(7.14)$$

$$h_1 + h_2 = h \qquad ...(7.15)$$

The solution of $h_1$ = 0.293h can be used to establish the coordinates of nodes 2 and 4.

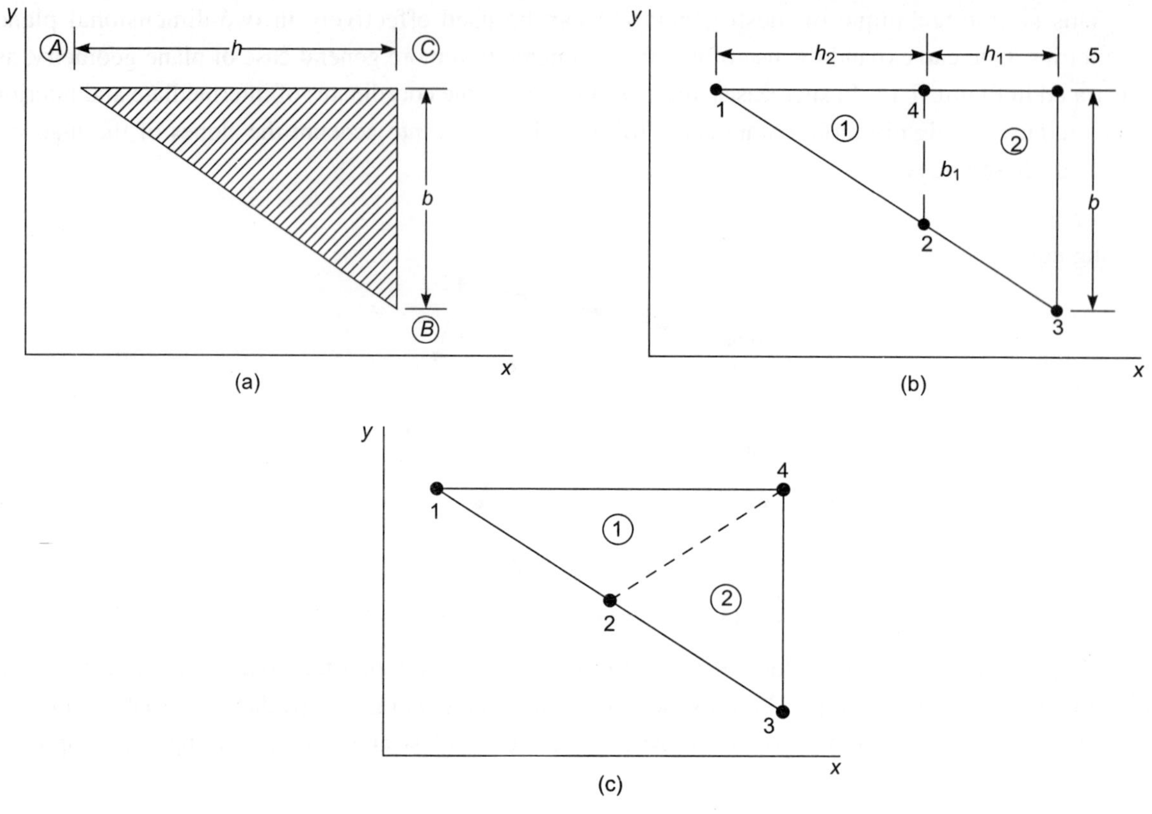


**Figure 7.14**

There are other ways in which this region can be discretized into two elements with equal areas, as illustrated in Figure 7.14c. The area of triangular plane elements with specified nodal coordinates of its vertices can be determined by the determinant of the matrix [*A*].

## Example 5

Determine the coordinates of node 2 in a region similar to what is shown in Figure 7.14c with the coordinates of the three vertices given as follows:

Vertex *A* (node 1): $x_1 = 2$ $y_1 = 8$ cm

Vertex *B* (node 3): $x_3 = 10$ $y_3 = 2$ cm

Vertex *C* (node 4): $x_4 = 2$ $y_4 = 8$ cm

The area of the region is computed to be $A = 30$ cm$^2$. If we let $A_1$ and $A_2$ be the respective areas for elements 1 and 2 and let the coordinates of node 2 be $(x, y)$, the areas of both elements can be determined by the determinant of the matrix [*A*] in Equation 7.5, or

$$2A_1 = (x_1y - xy_1) + (xy_4 - x_4y) + (x_4y_1 - x_1y_4)$$

$$2A_2 = (xy_3 - x_3y) + (x_3y_4 - x_4y_3) + (x_4y - xy_4)$$

The following two equations are obtained after substituting the numerical values of $x_1, y_1, \ldots x_4, y_4$ into the above expressions.

$$2A_1 = -10y + 80 \quad \ldots(7.16)$$

$$2A_2 = -6x + 2y + 56 \quad \ldots(7.17)$$

Since the area of element 1 is the same as the area of element 2, (i.e. $A_1 = A_2$), one can obtain the following equation by equating Equations 7.16 and 7.17.

$$6x - 12y = -24 \quad \ldots(7.18)$$

The condition of $A = 2A_1 = 2A_2 = 30\ \text{cm}^2$ can lead to the solution of $y = 5$ cm from Equation 7.16, and the solution of $x = 6$ can then be found from Equation 7.18 with the value of $y$. The coordinates of node 2 are thus (6, 5).

### 7.5.3 Mesh Generation by Volume Density Distribution

The concept of specifying the density distribution of element areas for two-dimensional finite element meshes can be extended to three-dimensional models. In such cases, the density distribution of element volume is used. The volume of the basic three-dimensional elements (i.e., tetrahedron elements, depicted in Figure 7.4 can be determined by the following determination:

$$V = \frac{1}{6}\begin{bmatrix} 1 & 1 & 1 & 1 \\ x_1 & x_2 & x_3 & x_4 \\ y_1 & y_2 & y_3 & y_4 \\ z_1 & z_2 & z_3 & z_4 \end{bmatrix} \quad \ldots(7.19)$$

where $(x_1, y_1, z_1)$, $(x_2, y_2, z_2)$, $(x_3, y_3, z_3)$ and $(x_4, y_4, z_4)$ are the coordinates of the four nodes of the element, as shown in Figure 7.15.

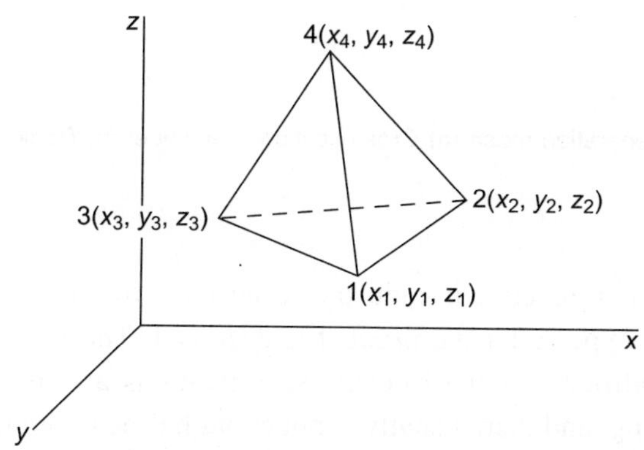


**Figure 7.15** Nodal coordinates of a tetrahedron element

This mesh generation scheme is illustrated by the following example involving the discretization of a wheel with a variable cross-section, as depicted in Figure 7.16a. The region for the finite element mesh can be approximated by a section of a pie shape with variable thickness, as shown in Figure 7.16b. One may first divide the region into two subvolumes by the plane designated by JKHG. The finite element mesh in each subvolume can then be generated by the users specification of the number of tetrahedron elements in each subvolume. We may visualize that the subvolume near the axle AD of the wheel can be subdivided into three tetrahedrons, ADKH, AGJK, and AGHK, as illustrated in Figure 7.16b. It is conceivable that a minimum of four tetrahedrons can be generated in the other subvolume. The locations, and thus the coordinates, of all intermediate nodes in the mesh can be determined by using the formula for the computation of the volume of a tetrahedron, given in Equation 7.19.

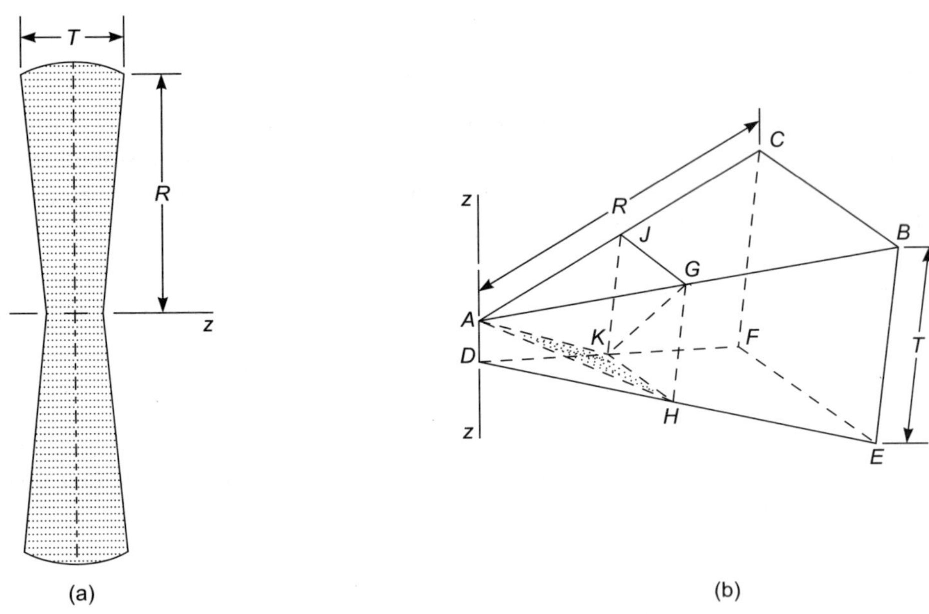


**Figure 7.16** A 3D mesh generation mesh (a) Cross-section of a wheel (b) Region for finite element model

## 7.6 CONCLUSIONS

Prototype building is a common practice in industry because the traditional manufacturing process for a new product requires prototypes after the product is designed. These prototypes allow engineers to conduct tests and make modifications until design specifications are met. However, this practice is expensive and time consuming, and there usually is not enough time to build these prototypes as design cycles are shortened by the competitive marketplace. Skipping this step in the manufacturing process is now possible with the sophisticated geometric modelling of the product and design analyses, which use such reliable techniques as the Finite Element Analysis and design optimization. Generally speaking, there are two ways of developing a finite element analysis. The first approach is to place emphasis on the accuracy of the result. This approach requires the development of sophisticated algorithms, such as high-order and special elements. Most of such work is tailored to some special purpose. The second approach is to develop general-purpose programmes and commercial codes such as ANSYS, NASTRAN, and MARC analysis. The seemingly unlimited capability of the FEA for handling design analyses of machine components having extremely complex geometries and load and boundary conditions makes it an important component in the CAD technology.

# CHAPTER 8

# Design for X

## 8.1 INTRODUCTION

Although the cost of design as such is a small fraction of the cost of product development (only about 8%), a major part of the product's cost (about 80%) is committed at the design stage (The concept or architecture of the product alone determines 60% of the cost!). The decisions arrived at this stage greatly influence the ease or hardship with which subsequent activities of product development and manufacture are carried out. Apart from these activities, the design decisions also influence other activities during the life of the product up to recycling it. Reorienting the design process so as to make these activities during the product development and product life cycles is referred as *Design for X* (*DFX*). Here, X refers to a number of factors such as

- Manufacture
- Assembly
- Mass Customization
- Safety
- Maintainability Serviceability
- Environment
- etc.

*Design for X* involve the designer with other functional departments, such as marketing, manufacturing, and engineering services, via Concurrent Engineering (CE) methods. In fact, DFX is often considered as a subset of CE. Since the other departments involve themselves at the early stages of design and help designers detect and correct flaws, such a CE environment is increasingly being referred as Collaborative Engineering.

DFX can be defined as a knowledge-based approach that attempts to design products that maximize all desirable characteristics — such as high-quality, reliability, serviceability, safety, user friendliness, environmental friendliness, and short-time-to-market in a product design while at the same time minimizing lifetime costs, including manufacturing costs.

Historically, designers have tended to underemphasize or overlook the preceding factors, and have concentrated their efforts on only three factors, viz., the *function* (performance), *features* and *appearance* of the product that they develop. They have tended to neglect the "downstream" considerations that affect the usability and cost of the product during its lifetime.

Henry Ford introduced many features which gave the Model T the competitive advantage. He designed the product so that it was easy to drive by virtually anyone. He designed the product to be repairable by any mechanic with tools readily available. He designed the parts of the product to be interchangeable and easy to attach to other parts. He ensured that interchangeable operators are available within the production system. He designed the movable production line. The combination of these sub-designs led to the integrated design of mass production, for which he is known. The salient point here is that, in order to capture the competitive advantage, he designed for many different characteristics. In other words, Henry Ford designed for X.

The three elements of competitiveness are:

(*i*) ***Timeliness***: Can the product be delivered to the market just in time to meet the customer's demand?

(*ii*) ***Quality***: Will the product perform well as intended?

(*iii*) ***Affordability***: Can customers afford to buy the product?

These three dimensions largely determine the value of a product to the customer. To gain the competitive advantage one must design for X. During design, one often focuses on the final product, and not its manufacture. DFX philosophy suggests that a design be continually reviewed from the start to the end to find ways to improve production and other aspects. These rules are nothing new, they are just common sense items written down, but they can be a good guide through the design process.

Advantages of these techniques are as follows:

- Shorter production times
- Fewer production steps
- Smaller parts inventory
- More standardized parts
- Simpler designs that are more likely to be robust
- They can help when expertise is not available, or as a way to reexamine traditional designs
- Proven to be very successful over decades of application.

In the subsequent sections, each facet of DFX will be reviewed.

## 8.2 DESIGN FOR MANUFACTURE

*Design for manufacture (DFM)* is the process of *proactively* designing products to

(*i*) Optimize all the manufacturing functions such as fabrication, assembly, test, procurement, shipping, delivery, service, and repair.

(*ii*) Assure the best cost, quality, reliability, regulatory compliance, safety, time-to-market, and customer satisfaction.

DFM is a proven design methodology that works for any size company. Early consideration of manufacturing issues shortens product development time, minimizes development cost, and ensures a smooth transition into production for quick time to market.

Quality can be built in with optimal part selection and proper integration of parts, for minimum interaction problems. By considering the cumulative effect of part quality on product quality, designers are encouraged to carefully specify part quality.

Many costs are reduced, since products can be quickly assembled from fewer parts. Thus, products are easier to build and assemble, in less time, with better quality. Parts are designed for ease of fabrication and commonality with other designs. DFM encourages standardization of parts, maximum use of purchased parts, modular design, and standard design features. Designers will save time and money by not having to "reinvent the wheel." The result is a broader product line that is responsive to customer needs.

Companies that have applied DFM have realized substantial benefits. Costs and time-to-market are often cut in half with significant improvements in quality, reliability, serviceability, product line breadth, delivery, customer acceptance and, in general, competitive posture.

In order to design for manufacture, everyone in product development team needs to,

- understand how products are manufactured through experience in manufacturing, training, rules/ guidelines, and/ or multi-functional design teams with manufacturing participation.
- specifically, design for the processes to be used to build the product he/ she is designing. If products will be built by standard processes, design teams must understand them and design for them. If processes are new, then design teams must concurrently design the new processes as they design the product.

Before DFM, "I designed it; you build it!" syndrome existed. Design engineers worked alone or only in the company of other design engineers in "The Engineering Department." Designs were then thrown over the wall leaving manufacturing people with the dilemma of either objecting (but its too late to change the design!) or struggling to launch a product that was not designed for manufacture (Figure 8.1). Often this delayed both the product launch and the time to ramp up to full production, which is the only meaningful measure of time-to-market.

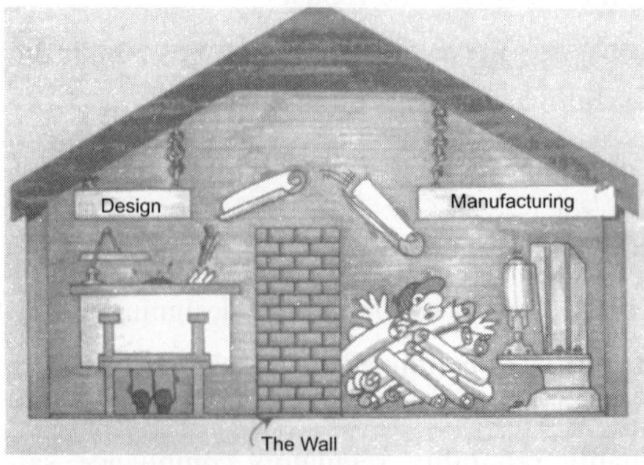


**Figure 8.1** Walls among functional groups in the factory – "I designed it; you build it!" syndrome

One way to assure manufacturability is by developing products in multi-functional teams with early and active participation from manufacturing, marketing (and even customers), finance, industrial designers, quality, service, purchasing, vendors, regulation compliance specialists, lawyers, and factory workers. The team works together to not only design for functionality, but also to optimize cost, delivery, quality, reliability, ease of assembly, testability, ease of service, shipping, human factors, styling, safety, customization, expandability, and various regulatory and environmental compliance.

Paradoxically, one of the first decisions the team has to make is the optimal use of off-the-shelf parts. In many cases, the architecture may have to literally be designed around the off-the-shelf components, but this can provide substantial benefits to the product and the product development process.

Off-the-shelf parts are less expensive to design considering the cost of design, documentation, prototyping, testing, the overhead cost of purchasing all the constituent parts, and the cost of non-core-competency manufacturing. Off-the-shelf parts save time considering the time to design, document, administer, and build, test, and fix prototype parts.

Suppliers of off-the-shelf parts are more efficient at their specialty, because they are more experienced on their products, continuously improve quality, have proven track records on reliability, design parts better for DFM, dedicate production facilities, produce parts at lower cost, offer standardized parts, and sometimes pick up warrantee/service costs. Finally, off-the-shelf part utilization helps internal resources focus on their *real* missions: designing products and building products.

### 8.2.1 Some Key DFM Guidelines

1. *Understand manufacturing problems/ issues of current/ past products*: In order to learn from the past and not repeat old mistakes, it is important to understand all problems and issues with current and past products with respect to manufacturability, introduction into production, quality, repairability, serviceability, regulatory test performance, and so on. This is especially true if previous engineering is being "leveraged" into new designs.
2. *Design for easy fabrication, processing, and assembly*: Designing for easy parts fabrication, material processing, and product assembly is a primary design consideration. Even if labor cost is reported to be a small percentage of the selling price, problems in fabrication, processing, and assembly can generate enormous costs, cause production delays, and demand the time of precious resources. Figure 8.2 shows two types of threads, one difficult to make and the other easy to make.

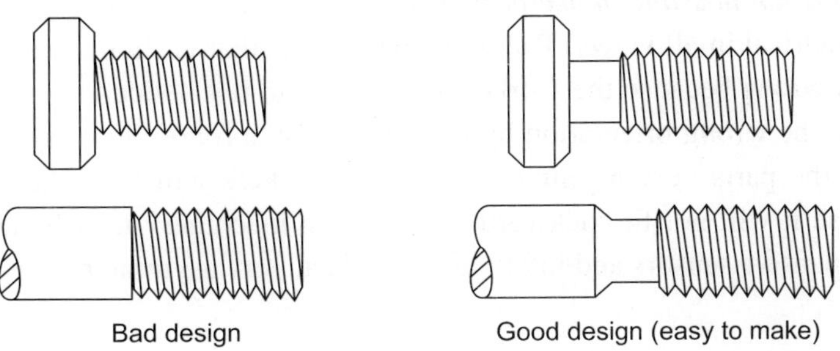


**Figure 8.2** Redesign of threads for ease of manufacture

3. *Adhere to specific process design guidelines*: It is very important to use specific design guidelines for parts to be produced by specific processes such as welding, casting, forging, extruding, forming, stamping, turning, milling, grinding, powdered metallurgy (sintering), plastic moulding, etc. Reference books are available that give a summary of design guidelines for many specific processes.
4. *Avoid right/ left hand parts*: Avoid designing mirror image (right or left hand) parts. Design the product so the same part can function in both right or left hand modes. If identical parts cannot perform both functions, add features to both right and left hand parts to make them the same. Another way of saying this is to use "paired" parts instead of right and left hand parts. Purchasing of paired parts (plus all the internal material supply functions) is for twice the quantity and half the number of types of parts. This can have a significant impact with many paired parts at high volume. At one time or another, everyone has opened a briefcase or suitcase upside down because the top looks like the bottom. The reason for this is that top and bottom are identical parts used in pairs.
5. *Design parts with symmetry*: Design each part to be symmetrical from every "view" (in a drafting sense) so that the part does not have to be oriented for assembly. In manual assembly, symmetrical parts cannot be installed backwards, a major potential quality problem associated with manual assembly. In automatic assembly, symmetrical parts do not require special sensors or mechanisms to orient them correctly. The extra cost of making the part symmetrical (the extra holes or whatever other feature is necessary) will probably be saved many times over by not having to develop complex orienting mechanisms and by avoiding quality problems. An example of converting an unsymmetric feature into symmetric one to simplify handling is shown in Figure 8.3.

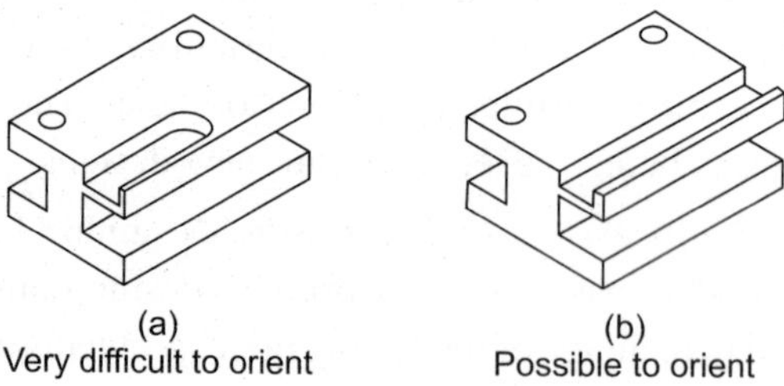


**Figure 8.3** Redesign to extend an unsymmetrical notch to a symmetrical one for ease of orientation

6. *If part symmetry is not possible, make parts very asymmetrical*: The best part for assembly is one that is symmetrical in all views. The worst part is one that is slightly asymmetrical which may be installed wrong because the worker or robot could not notice the asymmetry, the part may be forced in the wrong orientation by a worker or by a robot. So, if symmetry cannot be achieved, make the parts very asymmetrical. Then workers will less likely install the part backward because it will not fit backward. Automation machinery may be able to orient the part with less expensive sensors and intelligence. In fact, very asymmetrical parts may even be

able to be oriented by simple stationary guides over conveyor belts (Figure 8.4).

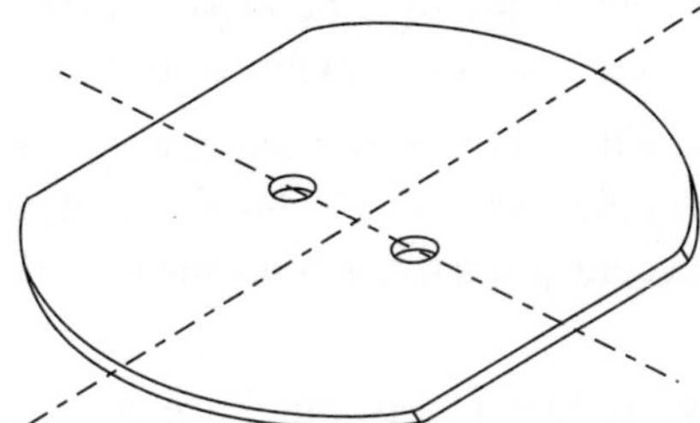

**Figure 8.4** Introduction of asymmetry to ensure correct orientation of an internal feature

7. *Design for fixturing*: Understand the manufacturing process well enough to be able to design parts and dimension them for fixturing. Parts designed for automation or mechanization need location registration features for fixturing. Machine tools, assembly stations, automatic transfers and automatic assembly equipment need to be able to grip or fixture the part in a known position for subsequent operations. This requires registration locations on which the part will be gripped or fixtured while part is being transferred, machined, processed or assembled.
8. *Minimize tooling complexity by concurrently designing tooling*: Use Concurrent Engineering of parts and tooling to minimize tooling complexity, cost, delivery lead-time and maximize throughput, quality and flexibility.
9. *Specify optimal tolerances for a Robust Design*: *Design of Experiments* can be used to determine the effect of variations in all tolerances on part or system quality. The result is that all tolerances can be optimized to provide a robust design to provide high quality at low cost. Tolerance also depends of the assembly process adopted. As shown in Figure 8.5, automatic assembly, manual assembly and selective assembly respectively are in the increasing order of tolerances.

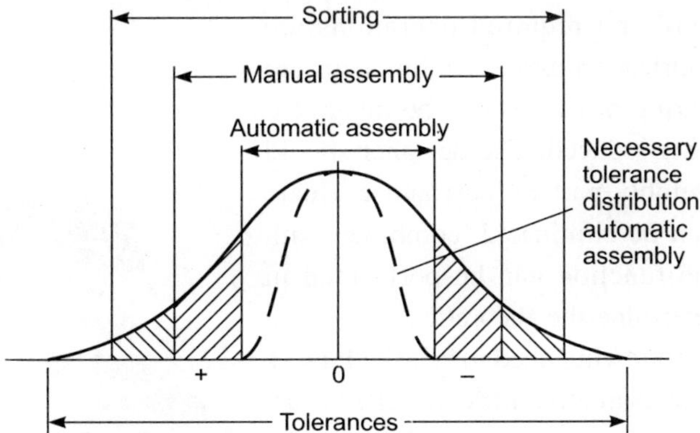


**Figure 8.5** Tolerances required for manual, automatic and selective modes of assembly

10. *Specify quality parts from reliable sources*: The "rule of ten" specifies that it costs 10 times more to find and repair a defect at the next stage of assembly. Thus, it costs 10 times more to

find a part defect at a sub-assembly; costs 10 times more to find a sub-assembly defect at final assembly; costs 10 times more in the distribution channel; and so on. All parts must have reliable sources that can deliver consistent quality over time in the volumes required.

11. *Minimize Setups*: For machined parts, ensure accuracy by designing parts and fixturing so all key dimensions are all cut in the same setup (chucking). Removing the part to reposition for subsequent cutting lowers accuracy relative to cuts made in the original position. Single setup machining is less expensive too.
12. *Minimize Cutting Tools*: For machined parts, minimize cost by designing parts to be machined with the minimum number of cutting tools. For material removal in CNC milling, specify radii that match the preferred cutting tools (avoid arbitrary decisions). Keep tool variety within the capability of the tool changer.
13. *Understand tolerance step functions and specify tolerances wisely*: The type of process depends on the tolerance. Each process has its practical "limit" to how close a tolerance could be held for a given skill level on the production line. If the tolerance is tighter than the limit, the next most precise (and expensive) process must be used. Designers must understand these "step functions" and know the tolerance limit for each process.

### 8.2.2 DFM Approach

1. *Simplify the design and reduce the number of parts*: Because for each part, there is an opportunity for a defective part and an assembly error, the probability of a perfect product goes down exponentially as the number of parts increases. As the number of parts goes up, the total cost of fabricating and assembling the product goes up. Automation becomes more difficult and more expensive when more parts are handled and processed. Costs related to purchasing, stocking, and servicing also go down as the number of parts are reduced. Inventory and work-in-process levels will go down with fewer parts. As the product structure and required operations are simplified, fewer fabrication and assembly steps are required, manufacturing processes can be integrated and lead-times further reduced. The designer should go through the assembly part by part and evaluate whether the part can be eliminated, combined with another part, or the function can be performed in another way. To determine the theoretical minimum number of parts, the following need to be asked: Does the part move relative to all other moving parts? Must the part absolutely be of a different material from the other parts? Must the part be different to allow possible disassembly? Figure 8.6 is an illustration of reducing an assembly into a single component by adopting casting process.

**Figure 8.6** Number of parts of an assembly being replaced by a casting

2. *Standardize and use common parts and materials*: This is required to facilitate design activities, to minimize the amount of inventory in the system, and to standardize handling and assembly operations. Common parts will result in lower inventories, reduced costs and higher quality. Operator learning is simplified and there is a greater opportunity for automation as the result of higher production volumes and operation standardization. Limit exotic or unique components because suppliers are less likely to compete on quality or cost for these components. Group technology (GT) and Component Supplier Management (CSM) systems can be utilized by designers to facilitate retrieval of similar designs and material catalogs or approved parts lists can serve as references for common purchased and stocked parts. GT can also be used to guide in the development of manufacturing cells for common part or product families, thereby minimizing inventory and providing improved effectiveness through manufacturing focus. For example, if different types of fasteners are avoided, a single spanner or screwdriver will be adequate (Figure 8.7). Another example is shown in Figure 8.8 where a number of toys have several common parts. This area of design methodology is also known as *Mass Customization*.

**Figure 8.7** Importance of standardization of fasteners

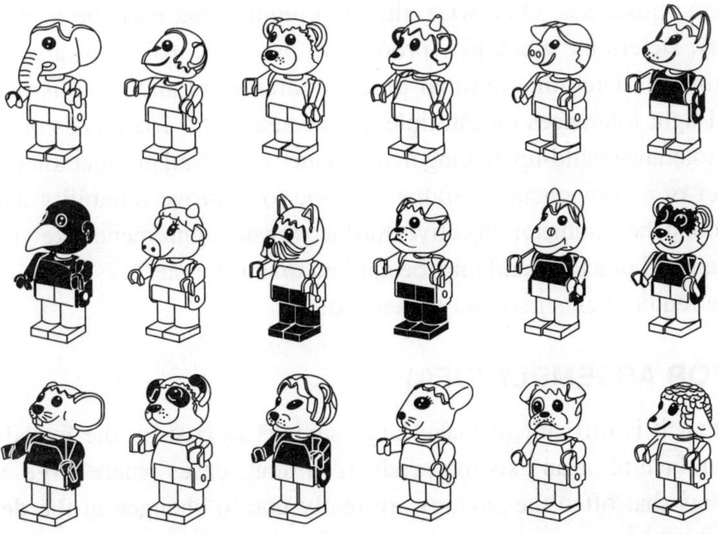

**Figure 8.8** Use of common parts in many products – Design for Mass Customization

3. *Design for ease of fabrication*: Select processes compatible with the materials and production volumes. Select materials compatible with production processes and that minimize processing time while meeting functional requirements. Avoid unnecessary part features because they involve extra processing effort and/ or more complex tooling. Consider specific guidelines appropriate for the fabrication process such as the following for machinability:
   - For higher volume parts, consider castings or stampings to reduce machining.
   - Use near net shapes for moulded and forged parts to minimize machining and processing effort.
   - Design for ease of fixturing by providing large solid mounting surface and parallel clamping surfaces.
   - Avoid designs requiring sharp corners or points in cutting tools — they break easier.
   - Avoid thin walls, thin webs, deep pockets or deep holes to withstand clamping and machining without distortion.
   - Avoid tapers and contours as much as possible in favour of rectangular shapes.
   - Avoid undercuts, which require special operations and tools.
   - Avoid hardened or difficult-to-machine materials unless essential to requirements.
   - Put machined surfaces on same plane or with same diameter to minimize number of operations.
   - Design work-pieces to use standard cutters, drill bit sizes or other tools.
   - Avoid small holes (drill bit breakage greater).
   - Avoid length to diameter ratio > 3 (chip clearance & straightness deviation).
4. *Design within process capabilities and avoid very high surface finish requirements (wherever possible)*: Know the production process capabilities of equipment and establish controlled processes. Avoid unnecessarily tight tolerances that are beyond the natural capability of the manufacturing processes. Otherwise, this will require that parts be inspected or screened for acceptability. Determine when new production process capabilities are needed early to allow sufficient time to determine optimal process parameters and establish a controlled process. Also, avoid tight tolerances on multiple, connected parts. Tolerances on connected parts will "stack-up" tolerance build-up making maintenance of overall product tolerance difficult. Design in the center of a component's parameter range to improve reliability and limit the range of variance around the parameter objective. Surface finish requirements likewise may be established based on standard practices and may be applied to interior surfaces resulting in additional costs where these requirements may not be needed.

## 8.3 DESIGN FOR ASSEMBLY (DFA)

Experience shows that it is difficult to make large savings in cost by the introduction of automatic assembly in the manufacture of an existing product. In those cases where large savings are claimed, examination will show that often the savings are really due to changes in the design of the product necessitated by the introduction of the new process. It can probably be stated that, in most of these instances, even greater savings would be made if the new product were to be assembled manually. Undoubtedly, the greatest cost savings are to be made by careful consideration of the design of the product and its individual component parts.

When a product is designed, consideration is generally given to the ease of manufacture of its individual parts and the function and appearance of the final product. Although for obvious reasons it must be possible to assemble the product, little thought is usually given to those aspects of design that will facilitate assembly of the parts and great reliance is often placed on the dexterity of the assembly operators. An operator is able to select, inspect, orient, transfer, place, and assemble the most complicated parts relatively easily, but many of these operations are difficult, if not possible, to duplicate on the machine. Thus, one of the first steps in the introduction of automation in the assembly process is to reconsider the design of the product so that the individual assembly operations become sufficiently simple for a machine to perform.

### 8.3.1 DFA Approach

1. *Design for ease of assembly*: Simple patterns of movement and minimizing the axes of assembly are recommended. Complex orientation and assembly movements in various directions should be avoided. Part features should be provided adequate chamfers and tapers (Figure 8.9). The product design should enable assembly to begin with a base component with a large relative mass and a low center of gravity upon which other parts are added. Assembly should proceed vertically with other parts added on top and positioned with the aid of gravity. This will minimize the need to re-orient the assembly and reduce the need for temporary fastening and more complex fixturing. A product that is easy to assemble manually will be easily assembled with automation. Assembly that is automated will be more uniform, more reliable, and of a higher quality.

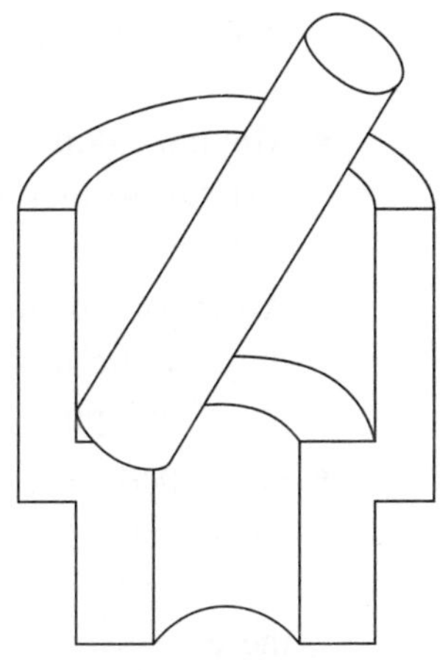

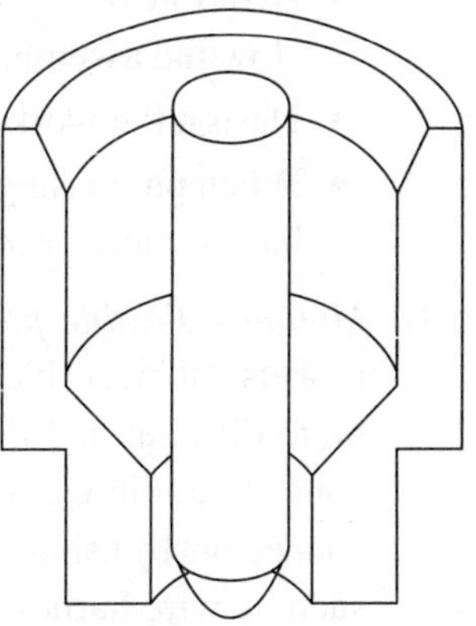

**Figure 8.9** Ease of assembly by introducing chamfers/ taper

2. *Fool-proof product design and assembly*: The assembly process should be unambiguous. Components should be designed so that they can only be assembled in one way; they cannot be reversed. Notches, asymmetrical holes and stops can be used to fool-proof the assembly process. Design verification of the product and its components is required. For mechanical products, this can be achieved with simple go/ no go tools in the form of notches or other features. Products should be designed to avoid adjustments. Electronic products can be designed to contain self-test and/ or diagnostic capabilities. Of course, the additional cost of building in diagnostics must be weighed against the advantages.

3. *Design for parts orientation and handling*: This is required to minimize non-value-added manual effort and ambiguity in orienting and merging parts. Basic principles to facilitate parts handling and orienting are:
   - Parts must be designed to consistently orient themselves when fed into a process.
   - Product design must avoid parts, which can get entangled, wedged or disoriented.
   - Avoid holes and tabs and use "closed" parts. This type of design will allow the use of automation in parts handling and assembly such as vibratory bowls, tubes, magazines, etc.
   - Part design should incorporate symmetry around both axes of insertion wherever possible. Where parts cannot be symmetrical, the asymmetry should be emphasized to assure correct insertion or easily identifiable feature should be provided.
   - With hidden features that require a particular orientation, provide an external feature or guide surface to correctly orient the part.
   - Guide surfaces should be provided to facilitate insertion.
   - Parts should be designed with surfaces so that they can be easily grasped, placed and fixtured. Ideally this means flat, parallel surfaces that would allow a part to picked-up by a person or a gripper with a pick and place robot and then easily fixtured.
   - Minimize thin, flat parts that are more difficult to pick up. Avoid very small parts that are difficult to pick-up or require a tool such as a tweezers to pick-up. This will increase handling and orientation time.
   - Avoid parts with sharp edges, burrs or points. These parts can injure workers or customers, they require more careful handling, they can damage product finishes, and they may be more susceptible to damage themselves if the sharp edge is an intended feature.
   - Avoid parts that can be easily damaged or broken.
   - Avoid parts that are sticky or slippery (thin oily plates, oily parts, adhesive backed parts, small plastic parts with smooth surfaces, etc.). Suitable dispensers can be used for applying these.
   - Avoid heavy parts that will increase worker fatigue, increase risk of worker injury, and slow the assembly process.
   - Design the workstation area to minimize the distance to access and move a part.
   - When purchasing components, consider acquiring materials already oriented in magazines, bands, tape, or strips.
4. *Minimize flexible parts and interconnections*: Avoid flexible and flimsy parts such as belts, gaskets, tubing, cables and wire harnesses. Their flexibility makes material handling and assembly more difficult and these parts are more susceptible to damage. Use plug-in boards and back-plates to minimize wire harnesses. Where harnesses are used, consider fool-proofing electrical connectors by using unique connectors to avoid connectors being misconnected. Interconnections such as wire harnesses, hydraulic lines, piping, etc. are expensive to fabricate, assemble and service. Partition the product to minimize interconnections between modules and co-locate related modules to minimize routing of interconnections.

5. *Design modular products*: This is required to facilitate assembly with building block components and sub-assemblies. This modular or building block design should minimize the number of part or assembly variants early in the manufacturing process while allowing for greater product variation late in the process during final assembly. This approach minimizes the total number of items to be manufactured, thereby reducing inventory and improving quality. Modules can be manufactured and tested before final assembly. The short final assembly lead-time can result in a wide variety of products being made to a customer's order in a short period of time without having to stock a significant level of inventory. Production of standard modules can be levelled and repetitive schedules established.
6. *Design for automated production*: Automated production involves less flexibility than manual production. The product must be designed in such way that it can be easily handled with automated equipment. There are two automation approaches: flexible robotic assembly and high speed automated assembly. Considerations with flexible robotic assembly are: design parts to utilize standard gripper and avoid gripper/ tool change, use self-locating parts, use simple parts presentation devices, and avoid the need to secure or clamp parts. Considerations with high speed automated assembly are: use a minimum of parts or standard parts for minimum of feeding bowls, etc., use closed parts (no projections, holes or slots) to avoid entangling, consider the potential for multi-axis assembly to speed the assembly cycle time, and use pre-oriented parts.
7. *Design for efficient joining and fastening*: Threaded fasteners (screws, bolts, nuts and washers) are time-consuming to assemble and difficult to automate. Where they must be used, standardize to minimize variety and use fasteners such as self-threading screws and captured washers. Consider the use of snap-together-fit (Figure 8.10). Evaluate other bonding techniques with adhesives. Match fastening techniques to materials, product functional requirements, and disassembly/ servicing requirements.

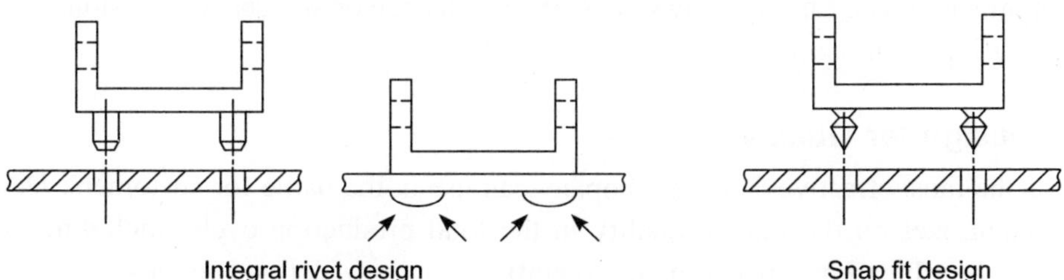


**Figure 8.10** Ease of assembly using snap fit

8. *Design printed circuit boards for assembly*: With printed circuit boards (PCB's), guidelines include: minimizing component variety, standardizing component packaging, using auto-insertable or placeable components, using a common component orientation and component placement to minimize soldering "shadows", selecting component and trace width that is within the

process capability, using appropriate pad and trace configuration and spacing to assure good solder-joints and avoid bridging, using standard board and panel sizes, using tooling holes, using minimum boards, and avoiding or minimizing adjustments.

## 8.4 OTHER FACETS OF DFX

DFM and DFA, together known as DFMA are the earliest and the most important facets of DFX. Some of its other facets are briefly described in this section.

### 8.4.1 Design for Environment

*Design for Environment (DFE)* is a systems level approach to product and process design where environmental attributes are treated as primary objectives or opportunities rather than as simple constraints. It is also known as *Green Engineering* or *Environmentally Conscious Manufacturing.*

It is an exciting new approach where organizations are forced to improve their financial performance without sacrificing their social obligation of protecting the environment. With the increasing environment conscience among people and promulgation of laws to protect it, organizations cannot afford to pollute the environment without incurring substantial penalty; at times violation may even lead to their closure.

In an increasingly competitive marketplace where price, quality and service are assumed, achieving real product differentiation is a key challenge. DFE offers a powerful way for companies to improve resource productivity and environmental performance and to win competitive advantage. DFE follows a life cycle thinking approach, which considers products and services from raw material extraction, through manufacture, use and final disposal. The many tools and examples now available mean that DFE can be utilized by all types of organization and integrated into mainstream product or process development.

Companies are today obliged to recycle a considerable portion of their products. The design has to focus more and more on materials that are bio-degradable. For instance, sponge is replaced by coir in many automobile seats. Emission laws have great influence on the design decisions, say on the type of fuels and its injection method.

### 8.4.2 Design for Quality

Quality is the most effective factor a company can use in the battle for customers. *Design for Quality (DFQ)* emphasizes on the role of quality in the total production cycle, including customer inputs, competitive benchmarking, performance specifications, product and process design, manufacturing variability and product reliability.

To be competitive, we must satisfy the customers. In order to be more competitive, we must delight the customer. Quality is defined here as the measure of *customer delight.* Note that customer satisfaction is a region on the scale of customer delightment. To delight the customer, we must design for quality.

*The domain of DFQ includes*: anticipating and satisfying customer expectations; a fundamental understanding of "variability" and the way it affects production processes; the new product life cycle and how to lower costs through merging design specifications and production; and more advanced topics such as robust design and the optimization of manufacturing processes.

*Kaizen*, a Japanese concept for continuous improvement, provides the philosophy and driving force for DFQ. *Total Quality Control* provides the implementation. The concepts are elegant. If quality is made the global driving force, then the customers will obtain the best value possible.

### 8.4.3 Design for Cost

Figure 8.11 illustrates the cost of various activities and their influence on the cost of the product. Effective product cost management requires a *Design for Cost (DFC)* philosophy as its basis since a

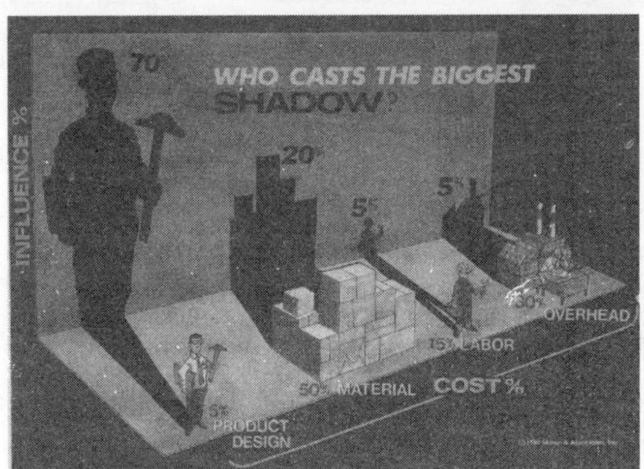


**Figure 8.11** Activity cost vs. Its influence on product cost

substantial portion of the product's cost is dictated by decisions at the design. Design for cost is a management strategy and supporting methodologies to achieve an affordable product by treating target cost as an independent design parameter that needs to be achieved during the development of a product. A design for cost approach consists of the following elements:

- An understanding of customer affordability or competitive pricing requirements by the key participants in the development process
- Establishment and allocation of target costs down to a level of the hardware where costs can be effectively managed
- Commitment by development personnel to development budgets and target costs
- Stability and management of requirements to balance requirements with affordability.
- An understanding of the product's cost drivers and consideration of cost drivers in establishing product specifications and in focusing attention on cost reduction
- Product cost models and life cycle cost models to project costs early in the development cycle to support decision-making
- Active consideration of costs during development as an important design parameter appropriately weighted with other decision parameters

- Creative exploration of concept and design alternatives as a basis for developing lower cost design approaches (Figure 8.12)

**Figure 8.12** Creativity and DFX

- Access to cost data to support this process and empower development team members
- Use of value analysis / function analysis and its derivatives (e.g., function analysis system technique) to understand essential product functions and to identify functions with a high cost to function ratio for further cost reduction
- Application of DFMA principles as a key cost reduction tactic
- Meaningful cost accounting systems using cost techniques such as *Activity-Based Costing (ABC)* to provide improved cost data
- Consistency of accounting methods between cost systems and product cost models as well as periodic validation of product cost models
- Continuous improvement through *Value Engineering* to improve product value over the longer term.

Figure 8.13 shows how some simple changes in the shape could lead to material saving.

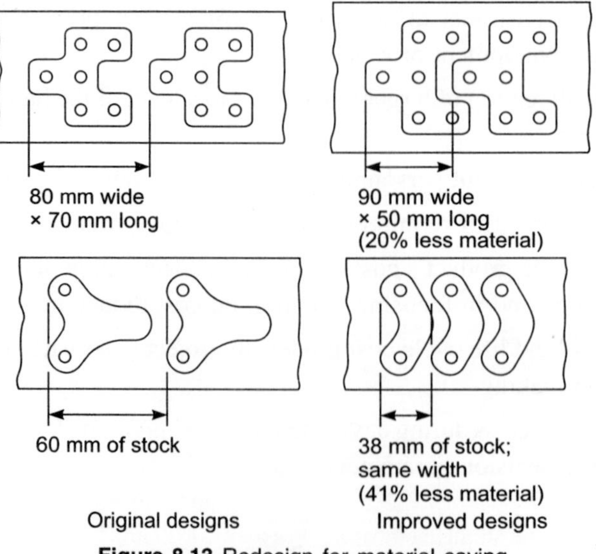


**Figure 8.13** Redesign for material saving

### 8.4.4 Design for Safety

*Design for Safety (DFS)* of an engineering system is a process of identifying the possible failure events and the associated consequences, estimating them and finally evaluating them. It provides the designer with a systematic approach to identify high-risk areas and attain explicit levels of safety by identifying and implementing ways to reduce the hazard frequency of occurrence and the extent of respective consequences. Risk identification and risk assessment are the most difficult and important steps that always attract a great deal of attention by safety researchers.

### 8.4.5 Design for Ergonomics

Most machines work in coordination with people. Human factors must be taken into account for every person who comes into contact with the product, whether during manufacture, operation, maintenance and repair, or disposal. The more the guessing required to understand the information and to control the action of the product, the lower the perceived quality of the product would be.

### 8.4.6 Design for Reliability

Reliability consideration has tended to be more of an after-thought in the development of many new products. Many companies' reliability activities have been performed primarily to satisfy internal procedures or customer requirements. Where reliability is actively considered in product design, it tends to be done relatively late in the development process. Some companies focus their efforts on developing reliability predictions when this effort instead could be better utilized understanding and mitigating failure modes, thereby developing improved product reliability. Organizations will go through repeated (and planned) design/ build/ test iterations to develop higher reliability products. Overall, this focus is reactive in nature, and the time pressures to bring a product to market limit the reliability improvements that might be made.

Specific Design for Reliability guidelines include the following:

- Design based on the expected range of the operating environment.
- Design to minimize or balance stresses and thermal loads and/or reduce sensitivity to these stresses or loads.
- Provide subsystem redundancy.
- Use proven component parts & materials with well-characterized reliability.
- Reduce parts count & interconnections (and their failure opportunities).
- Improve process capabilities to deliver more reliable components and assemblies.

### 8.4.7 Design for Maintainability/ Serviceability

Again, consideration of product maintainability/ serviceability tends to be an after-thought in the design of many products. Personnel responsible for maintenance and service need to be involved early to share their concerns and requirements. The design of the support processes needs to be developed in parallel with the design of the product. This can lead to lower overall life cycle costs and a product design that is optimized to its support processes. When designing for maintainability/ serviceability, there needs to be consideration of the trade-offs involved.

### 8.4.8 Design for Testability

If we can't test the product, how do we know how well it works? To ensure that a product meets the customer (user, owner, and regulator) desires, it must be tested for the ability to satisfy those desires. Only then do we know the value to the customer. In addition, testing can be lengthy and costly. In order to improve timeliness and cost, great care should be taken to design the product to be easy to test.

### 8.4.9 Other Important Facets of Xs

There are many other facets of DFX such as Design for Function & Performance, Design for Aesthetics, Design for Disassembly, Design for Features, Design for Short time–to-market, Design for Marketability, Design for Prototype, Design for Recycling, Design for Rapid Development, Design for Repair, Design for Reuse, Design for Simplicity, Design for Tolerance/ Location, Design for Portability, Design for Energy Conservation etc.

## 8.5 CONCLUSIONS

DFX, working in conjunction with Concurrent Engineering, helps in detecting and correcting design flaws early. It helps in evaluating the design not only for the functional requirements of the product but also for the ease with which the product can go through its life cycle smoothly. Although DFX sounds more of a philosophy, there are software and/ or detailed procedures laid out specific to some of its facets.

# Part III

## PHYSICAL PROTOTYPING AND RAPID MANUFACTURING

CHAPTER 9

# Computer Numerical Control

## 9.1 INTRODUCTION TO COMPUTER NUMERICAL CONTROL

*Computer Numerical Control (CNC)* is the latest machine tool control technique which enables high quality parts to be produced in small batches at low cost. Although special purpose machines, cam-operated automats, copying machines and transfer lines also can be used to produce high quality parts at low cost, their usage cannot be economically justified unless the quantity required is very high. This is because these are hard-automation equipment making use of cams, limit switches, relays and special wiring which are costly in terms of money and time and they have very poor reusable value when the company has to change from one product to the other or even when the same product is slightly modified. On the other hand, manufacturing systems built using CNC machines, robots, AGVs and other computer-controlled equipment are called flexible automation systems since switching over from one part or product to the other is achieved simply by changing the programmes (software) that control these equipment.

*Numerical Control (NC)*, as defined by Electronic Industries Association (EIA), is a machine tool control technique in which actions are controlled by the direct insertion of numerical data at some point. It is further stated that the system must automatically interpret at least some portion of this data. Although NC was first used for the control of machine tools of the metal cutting industry, today it is widely used for all kinds of conventional machining (milling machines, lathes, machining centers, boring machines, grinding centers etc.), Sheet metal manufacture (turret punch press, pipe bending machines, flow forming machines etc.), non-traditional machining (wire-cut EDM, water-jet cutting machines, plasma and laser cutting machines, electron beam welding machines etc.), inspection [*Coordinate Measuring Machines (CMM)*] and drafting (plotters).

In older NC machines, the control was by means of a micro-processor in which no RAM memory was available and the data was stored in the memory registers. Therefore, at any time only one NC block could be stored in the machine and many functions like interpolation, interlocking etc. were hard-wired. The NC programme was stored in paper tapes and the machine read only one block at a time from the paper tape, executed the machining instructions of that block, read the next block from the tape for execution and so on. Therefore, a paper tape reader had to be an integral part of the NC machine and it was continually read during machining. Due to the drastic and sustained fall in the prices of computer hardware, the NC machines manufactured during the last decade are each controlled by a dedicated micro-computer which has a large memory (RAM, EPROM and disk space in some cases) for storing NC programmes of several parts, it has a graphic monitor and also it is very fast. Because of this, the entire set of programmes to machine any part is read once and stored in its memory in a CNC machine and hence during the machining of the part, the tape reader is not used. Furthermore, many basic hard-wired functions of old NC machines are implemented as software for better flexibility and upgradability. The CNC machines are also able to perform additional functions to enhance the productivity of the machine tool. Some of them are graphic display of cutter path, conversational programming, machine tool diagnostics, storage of large number of cutter compensation values, axis calibration, adaptive control etc. Latest CNC controllers make use of multi-processors (transporters) or time-sharing techniques so that while the machining is going on, the operator can load or unload other programmes, edit the programmes etc. Since the old NC machines are obsolete, the terms 'NC' and 'CNC' are interchangeably used today which actually refers to CNC machines. A typical CNC machining center is shown in Figure 9.1.

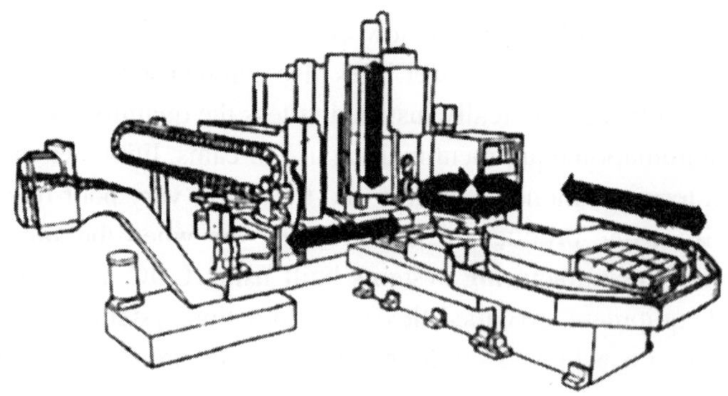

**Figure 9.1** A typical 5 axis CNC machining center

## 9.2 COMPARISON OF NC AND CONVENTIONAL MACHINES

As depicted in Figure 9.2, there are three major differences between a CNC machine and its conventional counterpart. They are

(i) Conventional machines need presence of operator

(ii) Metal removal rate is high in CNC machines

(iii) CNC machines have independent drives.

### 9.2.1 Presence of Operator

On a conventional machine, every movement required to machine the desired part is performed by the operator. He selects the spindle speed, feed rate and monitors them to compensate for slide friction, lack of power, and other changes in cutting conditions during machining. He centers the bar or locates the workpiece by trial and error and adjusts the clamps to avoid vibrations and deflections. Furthermore, he frequently measures the part at various stages of machining and controls the machine slides so as to

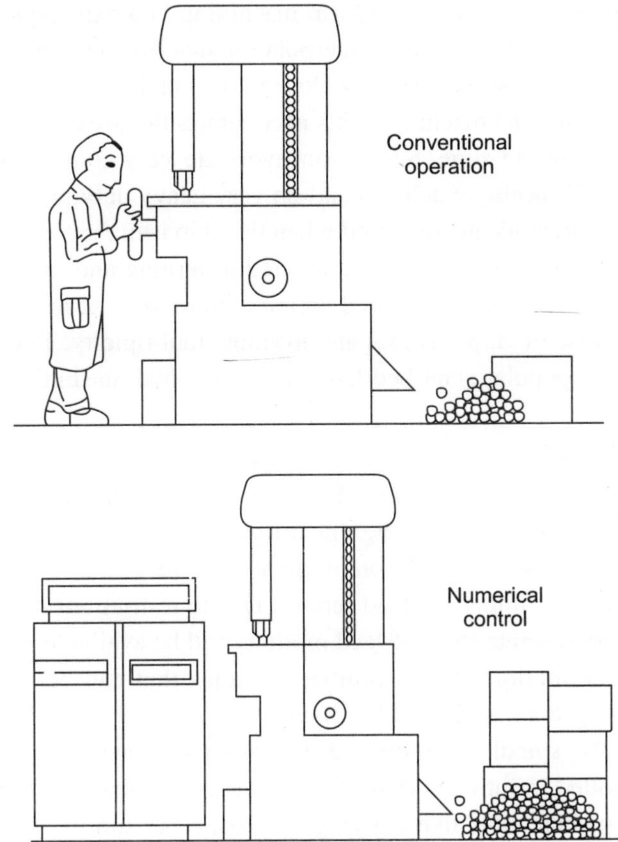


**Figure 9.2** Comparison of CNC and conventional machines

get the dimensions within the specified tolerance limits. It may be noted that very skilled operators can produce accurate parts even on old or inaccurate machines. On the other hand, on a CNC machine, the operator's job is limited to loading and unloading of the part and setting of the tools and all other minute machining steps are carried out automatically with the help of the NC programme. Since the operator cannot compensate for deficiencies like backlash, slide friction, lack of power etc., the machine tool design shall make sure that these deficiencies are as minimum as possible. Therefore, the CNC machine tool is built to be more accurate and rigid than its convention counterpart for the same application. We will see later some of the features that make a CNC machine more accurate.

### 9.2.2 Metal Removal Rate

On the CNC machines, the slides are power driven and hence the speed is only imited by the cutting speed allowable for the given combination of cutter and workpiece materials and the speed of the control system unlike the conventional machines where the cutting speed is limited. The feed and spindle motors are capable of producing the required power and owing to recent developments in cutting tool materials (carbide cutters, ceramic cutters and coated inserts), the cutting tools are able to withstand very high cutting speeds. In some free form machining in 5 axis mode especially of aircraft parts made of aluminum alloys, it is the speed of interpolation that limits the metal removal rate. Since the CNC machines in such application operate very close to the maximum possible, the metal removal rate is far higher than the conventional machines. This necessitates the use of fast chip disposal systems to remove the chips from the cutting zone. Screw conveyors are very commonly used chip conveyor mechanism on both milling and turning machines and on very heavy duty profilers, powerful vacuum extractors are used. Vacuum extractors are also used when the chip is in the form of fine power as in the case of grinding. Since continuous coil chips are generated in turning and drilling operations, special chip breaker design are required to be incorporated on the cutting tools. It has been found that radical design changes improve the ease of chip disposal and machine tool rigidity. Two examples that can be cited in support of this are the popular slant bed lathe and horizontal machining centers.

### 9.2.3 Independent Drives

In an ordinary lathe, there is only one motor which basically drives the spindle and the feed motions of the saddle is derived from the same motor through gear train and lead screw. In the same manner, the feed motions of all three slides, viz., vertical, longitudinal and cross slides of a conventional milling machine are obtained through gear trains and lead screws. Due to restrictions on the gear combinations that can be obtained, only very discrete steps of feed motions will be available in this method of drives. Furthermore, there will be limitation on the number of slides that can be simultaneously moved. Contrary to this, on a CNC machine, each axis has its own drive, drive control, measuring device and feedback loop. Therefore, the spindle rotation and the axes movement are all independent of each other and they can be controlled in infinite steps through the instructions in the NC programme. This avoids the requirement of special tools like copying devices, taper turning attachments and special form cutters. On a CNC machine, one can move the axes in linear or circular or any other higher order path with the help of interpolators built in the system. Therefore, even very complicated free form surfaces can be machined with greater ease without any special template or form cutter very accurately.

## 9.3 FEATURES OF A CNC MACHINE TOOL

Earlier we saw that a CNC machine tool is much more rigid and accurate than the conventional counterpart. CNC machines which have accuracy an of 0.008 mm and repeatability of a 0.004 mm are very common and CMM can measure features to an accuracy of 0.0002 mm. This high level of accuracy is achieved by improved mechanical and electrical hardware. Some of the features which help improve the accuracy of a CNC machine are:

- Rigid construction
- Anti-friction guideway elements
- Preloaded double-nut ball lead screws
- Advanced drives (stepper, dc servo and ac servo)
- Accurate position and velocity transducers

- Sophisticated control loops
- Axis calibration and pitch error compensation
- Compensation of deflections due to cutting forces and self-weight
- Control of room temperature
- Temperature compensation of the slide commands to take care of differences in the temperature at various points due to coolant flow etc.
- Accurate manufacture of the machine tool parts
- Diameter and length compensation.

The above features are built into the CNC machine tool by an appropriate design of its mechanical hardware, electrical and electronic hardware and computer hardware and software. The mechanical hardware consists of the machine tool structure, its guide ways and moving parts such as table and spindle head, power transmission elements such as gear trains, belts and clutches, power conversion elements such as lead screws or rack and pinion arrangements. The motors, limit switches etc. constitute the electrical hardware. The position and the velocity transducers are the major electronic hardware elements that may be digital, pulsed or analog devices. There are two types of computers present in any CNC machine. One is the CNC controller which takes care of input and output activities, decoding the NC blocks and filling appropriate word address registers, interpolation to calculate of command values for each axis etc. This is also known as *Data Processing Unit (DPU).* The other computer is called *Programmable Logic Controller (PLC).* This is responsible for interlocking of various switching functions by constantly monitoring the status of relays, switches and other switching devices. This also converts the command values into lower level voltage signal that will be processed by the drive unit. The PLC along with the drive control circuitry is sometimes known as *Control Loop Unit (CLU).* DPU and CLU together are called *Machine Control Unit (MCU).*

### 9.3.1 Mechanical Hardware

The machine tool is made very rigid and its moving parts are made as light as possible. The design of the structure shall be such that its natural frequency is well above that of the normally encountered disturbances to avoid resonance. In order to reduce friction, the machine tool guide way liners made of non-metallic material like Poly-Tetra Fluoro Ethylene (PTFE). In some cases, direct metal to metal contact of the moving parts is eliminated by using antifriction bearings, roller pads, hydrostatic, aerostatic and magnetic bearings. For very heavy machines, hydrostatic bearings are used. For machines which require very high accuracy and the load on it is fairly constant, aerostatic bearings minimize error due to thermal fluctuations. CNC machines with these bearings (also linear motors) can cut at 70 to 120 m/min. However, magnetic bearings used for CNC machines are still very costly.

Lead screws are used in small and medium size machines to convert the rotary motion of the shaft into the sliding motion of the table. The same is achieved by rack and pinion arrangement on larger machines. Conventional lead screws have ACME (trapezoidal) thread. These have sliding contact between the screw and nut surfaces. In the lead screws of CNC machines and accurate conventional machines such as jig borers and grinders, there is no direct contact of the sliding surfaces since balls are interspersed between them. Since these balls re-circulate and always a pair of nuts are used to reduce backlash, these lead screws are called "preloaded double nut re-circulating ball lead screws". The gap between the two nuts is optimally adjusted since higher gap will lead to more backlash whereas too narrow a gap will increase power consumption and wear. The adjustment of this gap is called 'preloading'. Figure 9.3 shows the details of a double-nut ball lead screw. Due to the rolling contact and continuous lubrication, these ball lead screws have considerably low friction. Furthermore, the difference between the static and dynamic friction coefficients is also very less and hence they also offer negligible 'stick-slip' behavior (erratic motion at low speed).

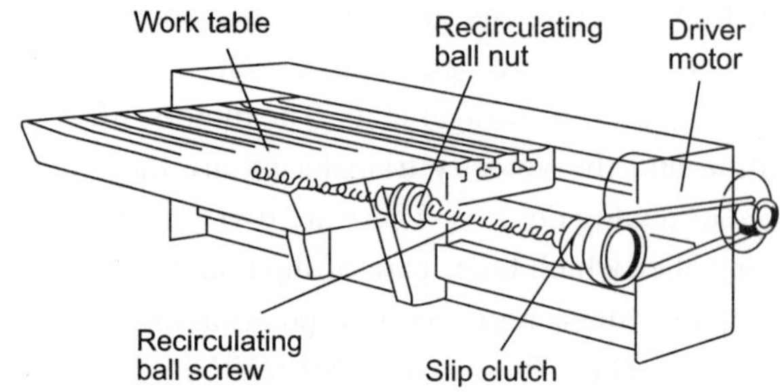


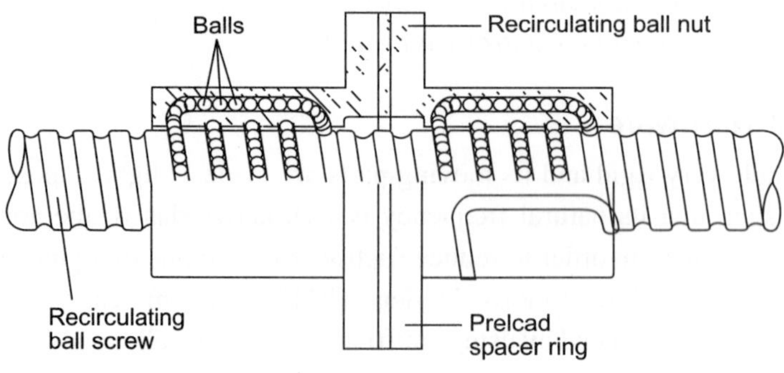


**Figure 9.3** Preloaded double-nut ball lead screw

### 9.3.2 Electrical and Electronic Hardware

The most common source of power for CNC machines is obtained from electric motors although pneumatic and hydraulic cylinders and motors are used in some specific applications. In the beginning, only stepper motors were used in CNC machines. A stepper motor moves by a fixed angle, say 7.5 degrees, for every pulse it receives. This movement is very accurate below a limiting speed and torque and hence stepper motors are generally used in open loop systems. However, when the cutter is overloaded beyond the limiting torque or when the commanded speed exceeds the limiting speed, it may fail to cope up leading to some lost motion. Therefore, today stepper motors are restricted to applications where the speed and torque are low and fairly constant such as floppy disc operations, plotter drives etc. Stepper motors are no longer used in CNC machines. They were initially replaced by DC servo motors and subsequently AC servo motors which work in closed loop control. The DC servo motor becomes noisy and its movements become jerky when the carbon brush wears depositing the powder on the commutator. Since the commutation is electronic rather than mechanical in AC servo motor, it is free from such maintenance problems. Both these servo motors work in conjunction with a velocity feed back element in a closed loop. The velocity feed back is obtained from a tacho-generator or an encoder mounted at the rear end of the motor shaft.

One is interested in controlling the target position, the path the cutter follows from the initial position and the rate of movement in CNC machines, transducers to measure the position and velocity are required. For incorporating adaptive control of machining, transducers for measuring force, torque, temperature etc. also will be required. Transducers may be rotary (tacho-generators, resolvers etc.) or linear (optical scale, inductosyn, magnascale etc.). When a rotary transducer is used to measure linear motion and vice versa, it is set to be an indirect transducer for that application. For example, the resolver mounted on the spindle motor itself can be used to calculate the table movements. Such an indirect measurement is less accurate than the direct measurement since it cannot reflect errors like pitch error, backlash etc. The signal produced by the transducer may be analog, pulses or digital. Figure 9.4 shows the optical scale used to measure linear and rotary positions.

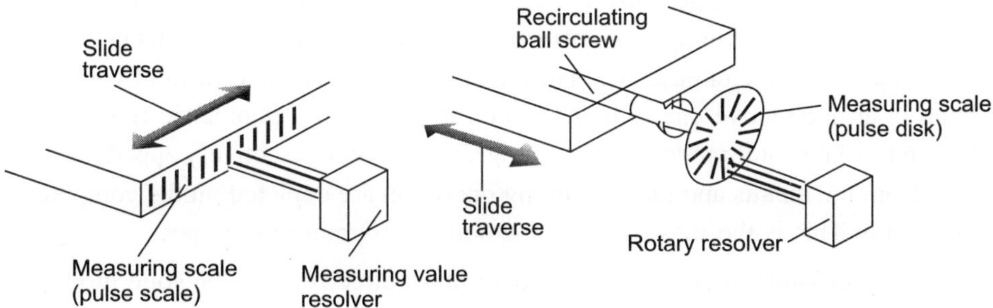


**Figure 9.4** Linear and rotary encoders used as feedback elements

## 9.4 PROGRAMMING CNC MACHINES

For each part, the user has to prepare a NC programme which is often called a part programme. A programmer who has very good understanding of the drawing requirements, Process planning, tooling and machining is responsible for preparing the part programme. The activity of part programming includes

- Reading the part print
- Deciding the process steps
- Deciding the tooling requirements

- Preparation of NC programmes
- Proving the programmes graphically, on soft material and finally on the actual material

The generation of NC programmes is done in the following three ways:

- Manual part programming
- Computer assisted part programming
- Automatic generation of NC programmes.

In manual part programming, the programmer calculates the path followed by the cutter center from the part geometry. He identifies the offset profiles in terms of lines and arcs and writes the NC blocks manually. An NC programme follows a word-address format in which an alphabet or a special character is followed by a number. This format is unique for each CNC controller. G and M are very commonly used codes. For example, G01 will move the cutter along a straight line path and G02 and G03 are used for circular motions. M03 and M04 are for "spindle on" and M05 is for "spindle off". Other words used are N, U, X, Y, Z, A, B, C, T, S, F, O etc. The programmer has to write the NC programme using these codes.

The manual part programming described earlier can be done for simple 2D shapes. When the shape is more complicated, it becomes difficult to perform the calculations manually. Therefore, the aid of computer is used. It is of two types: The first type is language based. The language used is called *Automatic Programming Tool (APT)* which is available in a variety of dialects such as APT-AC (from IBM), APT-140 (from British Aerospace), EXAPT, TC-APT etc. In the APT language, higher level commands like GOTO, TANTO (tangent to), PS (part surface), DS (drive surface), CS (check surface) etc. are available using which the programmer can write the machining steps and other control commands for spindle, coolant etc. with greater ease. The latest and more sophisticated form of computer assisted part programming is an interactive environment which is based on a geometric modeller (usually a surface modeller and of late solid modeller). The programmer can interactively specify the machining steps by selecting from the menus and the machining operation are depicted on the computer screen in the form of cutter path. This is the most popular method of programming today.

In future, it will be possible to generate NC programmes directly from the solid model of the part automatically, i.e. without any user interaction. Already this has been achieved for turned parts and limited success has been reported for milled parts also. In a CAD/CAM environment, the part geometry may be already available in the database as a solid model created by the designer. If this solid model is conventional B-Rep. or CSG type, then a feature extractor converts this model into a feature-based model. Alternatively, the designer himself can create the part geometry in terms of the manufacturing features like hole, slot, boss etc. An intelligent process planning system scans this modeller, determines the clamping schemes and cutter requirements for each set-up and then for each set-up, it also determines the cutter path for each cutter.

## 9.5 CONCLUSIONS

Important benefits that accrue out of using NC machines as compared to conventional manufacture are:

- Accuracy

- Flexibility
- Simplified tooling
- Reduction/ Elimination of set-up time
- Reduced machining time
- Ease of machining complicated contours
- Less skilled operators
- Less fatigue to the operators

Furthermore, NC machines are the basic building blocks for *Flexible Manufacturing Systems (FMS)* and *Computer-Integrated Manufacturing Systems (CIMS)*. It is the computer control of the manufacturing equipment that led to the integration of these equipment using networks. As a whole, NC plays a vital role in the conversion of Art (design) to Part (manufactured product).

❑❑❑

# CHAPTER 10

# Robotics

## 10.1 INTRODUCTION

Robots are re-programmable multi-functional mechanical manipulators. They can be programmed to perform a variety of jobs, which usually involve moving or handling objects. The following are the advantages of using robots:

- Flexibility
- High productivity
- Better quality
- Less risk to human life (as robots can be deployed in the hazardous jobs).

For a machine to qualify as a robot it must have the following five parts:

(i) Controller
(ii) Arm
(iii) Drive
(iv) End effector
(v) Sensors

### 10.1.1 Controller

Every robot is connected to a computer, which keeps the pieces of the arm working together. This computer is known as the controller (Figure 10.1). The controller also allows the machine to be networked to other systems so that it may work together with other machines/ processes or robots.

Robots today have controllers that are run by programmes. Almost all robots are entirely programmed by people, so they can do only what they are programmed for. In future, controllers with *artificial intelligence* (*AI*) could allow robots to think on their own, even programme themselves. This could make the robots more self reliant and independent.

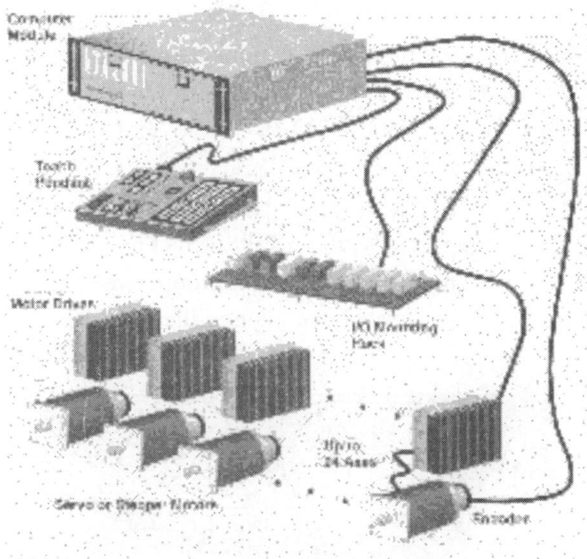

**Figure 10.1** Controller of a robot

## 10.1.2 Arm

Robot arms come in all shapes and sizes. The arm is the part of the robot that positions the end effectors and the sensors to do their preprogrammed operation (Figure 10.2). Many robots resemble human arms and have shoulders, elbows, wrists, even fingers. This gives the robot a number of ways to position itself in its environment. Each joint is said to give the robot one degree of freedom. So simple robotic arm with 3 degrees of freedom could move in three ways, viz., Up and Down, left and right, forward and backward. Most working robots have six degrees of freedom.

**Figure 10.2** A robotic arm

In order to reach any position in space with its work envelope, a robot needs a total of six degrees of freedom. As a result many robots of today are designed to move in at least six ways.

### 10.1.3 Drive

The drive is the "power" that drives the links (the sections between the joints) to their desired positions. Most drives are powered by air pressure, hydraulic fluids or electricity.

### 10.1.4 End Effector

The end effectors are the hands connected to the robots arm (Figure 10.3). It is often different from a human hand; it could be a tool such a gripper, a vacuum pump, tweezers, scalpel - just about any thing that helps in doing the final job. Some robots can change the end effectors and be reprogrammed for different set of tasks.

**Figure 10.3** End effectors of a robot

### 10.1.5 Sensors

Most of the robots today are nearly deaf and blind. Sensors can provide some limited feedback to the robot so it can do its job. Compared to the senses and abilities of even the simplest living things robots have a very long way to go.

The sensor reads and signals back to the controller. Sensors also give the robot controller information about its surroundings and let it know of exact position of the world around it. Sight, sound, touch, taste and smell are the other kinds of information we get from our world. Robots can be designed and programmed to get specific information that is beyond what our five senses can tell us; for instance, a robot sensor might see in the dark, detect tiny amount of movement that is too small or too fast the human eyes to see.

### 10.1.6 Robot Kinematics

The robot configurations are based on either prismatic or revolute joints whose position is described by a translation or rotation matrix respectively, measured by a feedback element such as a potentiometer or optical encoder. The servo control systems drive each joint to eliminate the error signal formed by subtracting the measured from the commanded position. The state of the robot is defined in joint space by the set of values representing each joint position. This may be different from actuator space if gear ratios vary or some drives are coupled. It is more convenient for the user to specify positions in natural coordinate systems such as world coordinates fixed with respect to the ground, or tool coordinates w.r.t. the gripper flange. In a Cartesian system, 3 translations specify the gripper position and also 3 rotations are needed to describe the gripper orientation, making 6 degrees of freedom in all. More may be needed to specify the state of the gripper itself. Fewer degrees of freedom are possible when some motions are excluded. Other standard names for frames include the base frame, station frame, wrist frame, tool frame and goal frame.

The geometrical relationships between these frames are a matter of kinematics, based on rigid links and close fitting joints. Whether the desired positions are in fact achieved will depend also on the dynamics taking into account drive torques, component inertia and speed.

## 10.2 CLASSIFICATION

Robots can be classified on the basis of the following parameters:

- Arm configuration
- Shape of the work space
- Operating method
- Type of motion
- Type of power
- Size
- Type and number of joints
- Type of technology
- Generation of design
- Type of motion

### 10.2.1 Classification by Arm Configuration

The following are some of the kinematic configurations of the robot arm:

- Cartesian robot
- Cylindrical robot
- Spherical robot
- Jointed-arm or Articulated robot
- Hexapod robot

Cartesian robots move along x, y, and z-axes in straight lines. Cylindrical robot manipulator can rotate about its base and can move linearly in horizontal and vertical direction. *SCARA* (*Selective Compliance Articulated Robotic Arm*) robot is similar to the cylindrical robot but it has a rotational reach axis rather than a linear one. Polar or Spherical manipulator can rotate about its head as well as its base. It can also move in and out. Jointed Arm or articulated robot arms resemble the design of a human arm and have largest work cell amount of required floor space. Hexapod robots have very high stiffness. For the same reason, this configuration is used even for CNC machines.

### 10.2.2 Classification by Type of Motion

- *Limited sequence*: It uses mechanical stops to limit the movement. A timer may be used to activate an axis.
- *Point to Point*: These types of controllers have sensors for feedback and memory to hold the coordinates of each axis. The actuators all run simultaneously at one speed, usually maximum. Some joints finish before others. The tool path is predictable, but not readily visualized.
- *Continuous Path*: These types of sensors have greater memories than point to point and can record many coordinates per second. The controller generates a sequence of interpolated via-points. This is one way of achieving joint-interpolated movement, but if the computer's arithmetic is fast enough, the points can be calculated to lie on a straight line or circular arc. A large number of inverse transformations must be performed. This method is also used after auto-recording of a sequence of points during lead-through teaching.
- *Joint-interpolated motion*: All joints start and finish together, being driven at speeds calculated from that of the slowest joint (or the one with farthest to travel) gives smoother movement, which is often closer, to what is expected, especially if intermediate via-points are included to constrain the departure from the desired track. There can be considerable variation from straight line motion, except where the robot's geometry permits natural movement in one coordinate plane or axis.

### 10.2.3 Classification by Power Supply

The power sources for robot actuation could be

- Electrical
- Pneumatic
- Hydraulic.

Electric robots are efficient, require little maintenance, and are not very noisy. Pneumatic robots use compressed air and come in a wide variety of sizes. A pneumatic robot requires another source of energy such as electricity, propane or gasoline to provide the compressed air. Hydraulic robots use oil under pressure and generally perform heavy duty jobs. This power type is noisy, large and heavier than the other power sources. A hydraulic robot also needs another source of energy to move the fluids through its components. Pneumatic and hydraulic robots require maintenance of the tubes, fittings and hoses that connect the components and distribute the energy.

Two important sources of electric power for mobile robots are solar cells and batteries. There are many types of batteries like carbon-zinc, lithium-ion, lead-acid, nickel-cadmium, nickel-hydrogen, silver-zinc and alkaline to name a few. Battery power is measured in amp. hours which is the current (amp) multiplied by the time in hours that current is flowing from the battery. For example, a two amp. hour battery can supply 2 amps. of current for one hour. Solar cells make electrical power from sunlight. If you hook enough solar cells together in a solar panel you can generate enough power to run a robot. Solar cells are also used as a power source to recharge batteries.

Deep space probes must use alternate power sources because beyond Mars existing solar arrays would have to be so large as to be infeasible. The lifespan of batteries is exceeded at these distances also. Power for deep space probes is traditionally generated by *radioisotope thermoelectric generators (RTG)*, which use heat from the natural decay of plutonium to generate direct current electricity. RTGs have been used on 25 space missions including Cassini, Galileo, and Ulysses.

### 10.2.4 Classification by the Level of Technology

- *Low Level Technology*: These robots are used for material handling, loading, and unloading, simple assembly operations. These machines have 2–4 axes.
- *Medium level technology*: These robots are used for picking and placing, loading and unloading, have 4–6 axes of movement.
- *High Level Technology*: These robots can be used for manufacturing tasks as well as for other engineering applications. These have 6–9 axes of movement and possess higher level of sensory perception.

### 10.2.5 Classification by Design

- *First generation*: These robots use fixed sequence of programmes. Reprogramming had to take place before another task could take place. They did not posses sensors and all control was through an open loop controller system.
- *$1.5^{th}$ generation*: These robots can adapt to varying situations by the use of sensors and a closed loop servo controller. Although the robot may know when something is wrong in the system it may not be capable of doing anything about it. It requests human help. These are today's robots.
- *Second generation*: These robots are in near future. Vision system in conjunction with better mobility systems will make the robots much more efficient and flexible. More intelligent microprocessors coupled with voice recognition and hand-to-hand coordination will expand the usefulness of the robot systems. These robots will be able to make minor decisions and will be able to make minor corrections or adjustments to manipulators.
- *$2.5^{th}$ generation*: Perceptual motor functions will be the highlight of these robots. Response to sensory stimuli will produce control motion. If initial sequence of steps are provided for robot task, the robot will be able to decide what to do next to accomplish the specific task.
- *Third generation*: These are really smart robots. If the robots have generalized information about what tasks need to be done, they will be able to decide the best way to accomplish the task.

### 10.2.6 LERT (Motion) Classification System

LERT classification system uses the following four motions as the basis for the classification, LERT being an acronym obtained by the first letters of these motions:

- *Linear motion*: This type of motion usually occurs when one part moves along the outside of another part such as in a rack and pinion system.
- *Extensional motion*: This is a type of telescoping motion produced when one part moves within another part. Kinematically this is same as linear type although constructionally different.
- *Rotational motion*: This type of motion is caused by a part, moving about something other than its centre.
- *Twisting motion*: Twisting motion is produced by a part turning about its centre.

LERT number for some of the robot configurations are as follows:

- Cartesian (only in front of itself) : L3
- Cylindrical (around itself) : L2R1
- Polar (above, below and around) : L1R2
- SCARA (around itself or around an obstacle) : R3

## 10.3 PROGRAMMING

Robots and CNC machines require one control programme for each job they perform. Whenever they switch from one job to the other, this control programme needs to be changed, except autonomous robots which can take their own decisions based on the changing situation as they possess sensors and are intelligent. These control programmes are a series of code that indicate the motions and switching functions they have to perform.

There are two ways in which these control programmes can be generated:

(i) Online programming
(ii) Offline programming.

Online programming is done using the robot or CNC machine itself. This is the easiest to do when the geometry and kinematics involved are simple. However, this employs the costly equipment unnecessarily and requires careful testing before actual use. On the other hand, offline programming is carried out on another computer which is invariably cheaper than the robot. Furthermore, it requires very less input from the programmer and does all the necessary calculations by itself. In the past, programming languages which were mostly dialects of *Automatically Programmable Tool (APT)* language were used for this purpose. Nowadays, geometric model based interactive programming is used instead of these languages which is more elegant. There are efforts to generate these programmes automatically, i.e., without any human intervention.

Different types of techniques used for online programming robots are:

- *Walk through or teach pendant programming*: The robot is moved using a hand held pendant that has buttons and/ or joysticks for various motions. The axes values at different positions of the robot obtained using the pendent are stored which can be played back. This is the cheapest way of programming for robots. Coordinate Measuring Machines (CMMs) also make use of this method of programming. However, this is not a very elegant and easy way of programming. It takes a lot of time to learn this method of programming.

- *Lead through programming*: In this method, the operator simply holds the arm of the robot and takes it along to the required path. During this motion, the positions are recorded at regular time intervals. Typically for paint spraying, this is the most preferred and popular method of programming industrial robot.
- *Tele-operation*: This is based on master-slave control. The actual robot (slave) and its replica (master) are kept in two different places and these communicate using cable or wireless means. Whenever the master is moved by the operator, the slave will faithfully follow it. The operator always needs to look at the slave robot through a glass window or with the help of video cameras. The limitation of this method is the need of the operator as long as the robot is in operation. This is used in hazardous environments such as nuclear installations, underwater, space etc., where the conditions are not predictable and hence a standard programme cannot be used. For non-engineering applications, such as robots for handicapped persons, robots tele-operated using voice are possible.

The programmes generated using above (except tele-operation where there is no programme at all) need to be validated to ensure that they produce parts that conform to design requirements and they are safe to use. This can be done online or offline. Solid modelling based offline simulations are becoming popular for validating the programmes of both robot and CNC machines.

## 10.4 SENSORS

Robots were initially expected to become humanoids. However, nonavailability of suitable sensors and the adequate hardware and software to process the sensory information at the required speed were the bottlenecks. Therefore, robots with less sensory perception could be used only for industrial applications. Developments in electronics and computer technology are overcoming these bottlenecks and more smarter and intelligent robots will be available in the near future.

The sensors for robotics can be classified as:

(i) Internal sensors

(ii) External sensors.

### 10.4.1 Internal Sensors

These sensors provide feedback signals for positional control of joints, except when stepping motors are used. Different types of internal sensors are:

- *Potentiometers*: These provide analogue signals, are unreliable because of track wear and are inaccurate, e.g. 0.1% linearity error represents 0.3 in 300, but this gives 2.6 mm at 0.5 m radius.

- *Optical encoders*: These provide digital signals easily, interfaced to a microprocessor. The incremental form uses a photocell to detect light interrupted by a series of obscured sectors of a disk, which rotates with the joint. Using two photocells, this simple counter can be used to detect direction too. An absolute encoder uses multiple tracks and photocells to provide (say) 15 bits of positional information. The sectors are patterned to provide a Gray code signal: this is more reliable than a binary code because only one bit changes from one sector to the next, eliminating spurious intermediate readings. High accuracy with no contact. Typical resolution is $360/2^{15}$ = 0.66 minutes of arc. Using a hybrid system with both a potentiometer and an encoder geared at 4:1 can increase resolution to 10 seconds of arc. Such accuracy demands precision manufacture, using anti backlash gearing. Only light springs are needed, as the drive torque for the encoder is low.
- *Linear Variable Differential Transformer (LVDT)*: This is a device that works on the principle of flux linkage. It can measure the position of a body at any time and so it can be used as a sensor. If used with differential *operational amplifier (opamp),* it can also give the velocity and acceleration of a body.

### 10.4.2. External Sensors

#### *10.4.2.1 Vision*

- *TV camera*: This requires fast *analog-to-digital converter (ADC)* and frame store to capture a digitized image. 6 or 8 bit ADC gives 64 or 256 gray levels for each pixel. It collects data at a very high rate (e.g. 500 points, 625 lines, 25 frames/sec, 8 bits/pixel = 62.5MHz bit rate) and has large storage requirements (e.g. 625 lines, 500 points = 300 kbytes per frame store).
- *CCD array*: Solid state charge-coupled device (CCD) cameras are smaller and lighter, suitable for mounting on the robot. Although the pixels are captured in an array (perhaps 256 × 400 or 512 × 512 resolution), parallei ADC channels would be needed to store a digital representation directly. Instead, shift registers transfer pixels out as a standard video signal, to be captured as before using an expensive frame grabber.
- *DRAM solid state cameras*: It uses the optical properties of dynamic RAM chips to capture binary images. The discharge rate of the individual capacitors forming the memory array is light sensitive. Using standard microprocessor memory interfacing chips allow the data to be accessed conveniently using cheap components. Different black/white threshold levels can be set by varying the discharge time after a refresh pulse. It permits image processing of 256 × 256 pixels on a standard computer.

Typically, we wish to identify a part and measure its position and orientation. Given isolated images of a number of non-overlapping components, the vision system might first construct a simple binary representation by coalescing similar grey levels, contrast enhancement, and by eliminating isolated pixels of one colour, noise reduction. Edge detection uses sharp gradients to identify component boundaries.

Simple operations on pixels of one colour then calculate recognition parameters such as:

- Connectivity (number of holes)
- Area and perimeter(s)
- Centroid (x and y values)
- Second moment of area
- Principal axes (lengths and directions)

Each detected object can be analyzed and its features compared with a standard set describing of each component to be recognized. Once identified, the centroid fixes the location and the angle of the major axis fixes the orientation, after which the robot can be programmed to approach the part and grasp it correctly.

#### *10.4.2.2 Tactile Sensing*

It uses rubber membrane above acrylic sheet with air gap in between. Sidelight normally contained by total internal reflection within the acrylic, but a CCD array beneath detects light scattered where the rubber is pushed down. The CCD provides an array of 'tactels', allowing tactile fingers to 'read' surface texture or detail.

Touch probe, a single-point probe connected to a variable resistance can determine component shape, or departure from a standard shape.

## 10.5 APPLICATIONS

The following are some of the applications of robotics:

### Industrial Applications

- Welding (arc welding, spot welding, seam welding and soldering)
- Material handling (Loading and unloading between machines and stores)
- Drilling and deburring
- Assembly
- Inspection
- Painting
- Packaging/ palletizing
- Underwater repairs
- Applications in nuclear installations
- Space applications

### Other Applications

- Biorobotics

- Medical applications
- Assistant to handicapped persons
- Cutting wool from sheep
- Deboning of meat
- Unmanned petrol bunks

Some of these applications are discussed in the following paragraphs.

## Welding

One of the common uses for industrial robots is welding. Robot welded car bodies have better safety. A robot (Figure 10.4) never misses a welding spot and performs equally all through the day. Nearly, 25% of the all industrial robots are used in different welding operations. Robot controls different welding parameters like wire feed and power. Seam welding robots have 5–6 axes since two and sometimes three axes are necessary for proper entry and continued operation of welds head to seam. These robots position a welding head in relation to the seam by either a trajectory method or seam sensing method. Trajectory method is simply programming a trajectory path that the robot will follow repeatedly, regardless of seam locations. Seam sensing is accomplished by either tracking the seam with a mechanical finger or by significantly costly vision system method. It is useful to consider robotic seam welding in situation where manual approach requires frequent operator movement, frequent work piece positioning or both to complete the welding operation.

**Figure 10.4** A robot undertaking a welding operation

## Assembly

Assembly accounts for nearly 33% of the industrial robots. Many of these robots (Figure 10.5) can be found in automotive and electronic industries. In addition to the robot, robotic assembly cell may require other equipment such as vision system. From a cycle time stand point, a manual assembly is sometimes faster than the robotic assembly, however, robotic assembly is often a good alternative in clean rooms and other environments where human presence is not desirable.

**Figure 10.5** A robot assembling car parts

## Painting

These robots (also known as finishing robots) are five or six axis units as shown in Figure 10.6. It has the capability of painting over a complex surface at a constant velocity relative to the surface.

**Figure 10.6** A painting robot

## Biorobotics

Recently a system of micro propagation of plants has been developed. The process involves harvesting each arriving plants, cutting its stem into segments near each node and then replanting the segments so that they can grow into new plants. The whole process is undertaken by the help of robots (Figure 10.7).

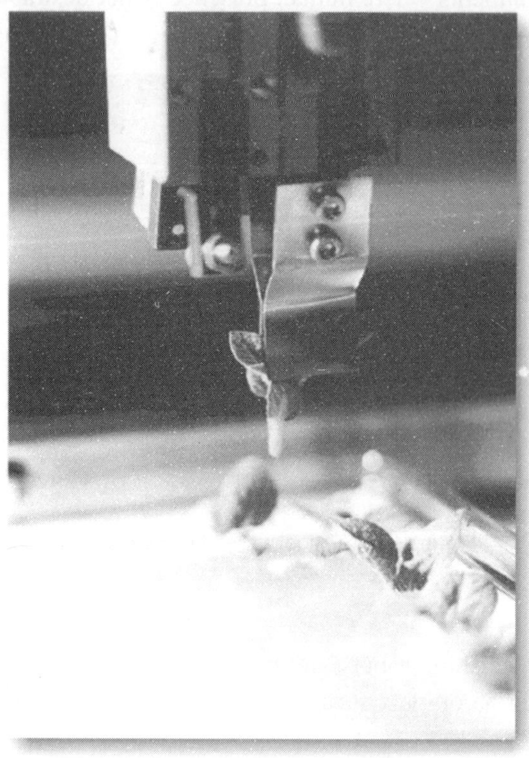

**Figure 10.7** A robot cutting the stems into segments near each node of the plant

## Medical Applications

*Minimally Invasive Surgery (MIS)* is a revolutionary approach in surgery. In MIS, an incision is made in the human body and a robot (Figure 10.8) is inserted into the body in fragments. The different parts of robots align themselves inside the human body and perform the required operations. Unfortunately, there are disadvantages due to reduced dexterity and sensory input to the surgeon, which is only available through a single image.

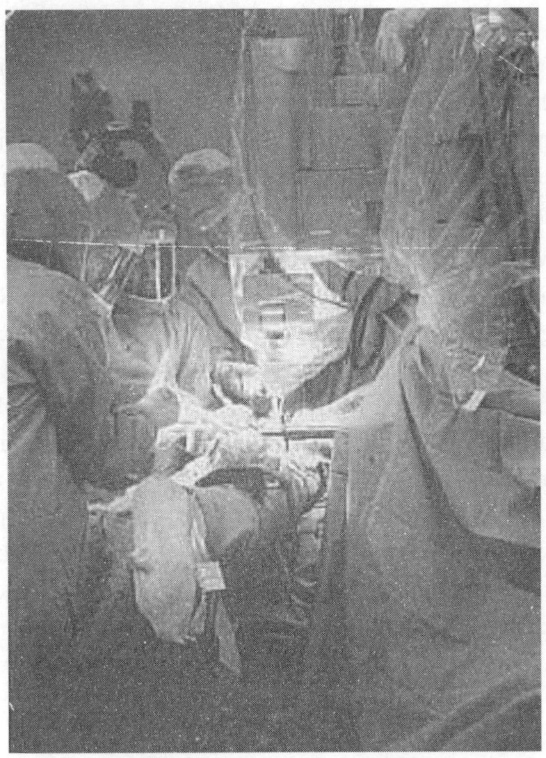

**Figure 10.8** Minimally invasive surgery

## Underwater Applications

Conventional unmanned, submersible robots are used in science and industry throughout the oceans of the world. To get around, automated underwater vehicles (AUV's) use propellers and rudders to control their direction of travel. An underwater robot (Figure 10.9) could propel itself as a fish does using its natural undulatory motion. It is thought that robots that move like fish would be quieter, more manoeuvrable and more energy efficient.

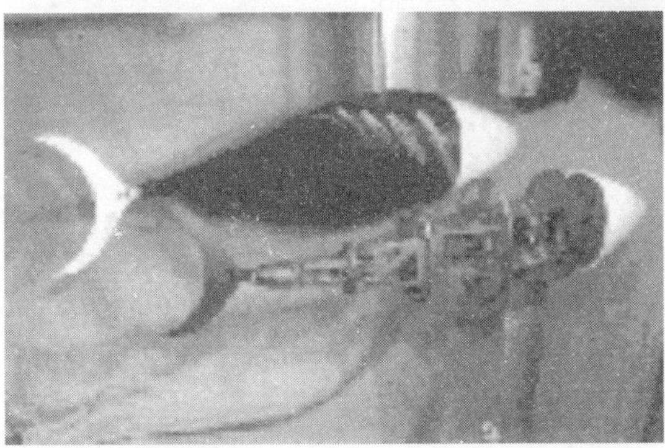

**Figure 10.9** An automated underwater vehicle

## Space Expeditions

Space-based robotic technology at NASA falls within three specific mission areas: (i) exploration robotics, (ii) science payload maintenance and (iii) on-orbit service. Related elements are terrestrial/ commercial applications, which transfer technologies generated from space telerobotics to the commercial sector, and component technology, which encompasses the development of joint designs, muscle wire, exoskeletons, and sensor technology.

Today, two important devices exist which are proven space robots. One is the *Remotely Operated Vehicle (ROV)* and the other is the *Remote Manipulator System (RMS)*. An ROV can be an unmanned spacecraft that remains in flight, a lander that makes contact with an extra terrestrial body and operates from a stationary position, or a rover that can move over terrain once it has landed. It is difficult to say exactly when early spacecraft evolved from simple automatons to robot explorers or ROVs. Even the earliest and simplest spacecraft operated with some preprogrammed functions monitored closely from Earth. One of the best known ROV's is the Sojourner (Figure 10.10) rover that was deployed by the Mars Pathfinder spacecraft. Several NASA centers are involved in developing planetary explorers, robotic technology, and space-based robots. To date, the NASA RMS robot arm has performed a number of tasks on many space missions, serving as a grappler, a remote assembly device, and also as a positioning and anchoring device for astronauts working in space.

**Figure 10.10** Sojourner on Mars

## Packaging/palletizing

This industry is still a minor application area and is expected to develop in coming years (Figure10.11). The food industry is a major area where robots are expected to play a major role in years to come (Figure 10.12).

**Figure 10.11** Robots in packaging

**Figure10.12** Robots in food industry

## 10.6 CONCLUSIONS

Although robots and CNC machines are similar in many ways, robots are less popular till recently in the industries than CNC machines. However, recent developments in sensors and signal and image processing have enhanced their capabilities substantially. Robots are now seen extensively in industries and soon they will be seen as an integral part of every household. These machines are here to stay because they are:

- Adaptable to problems and environments.
- Capable of sensing a wide range of sensory inputs, with pattern recognition.
- Able to make decisions, set priorities and define goals.
- Able to investigate new experimental techniques.
- Easy to programme.

But at the same time there remain certain social issues that need to be answered today since there is a fear that the use of robot will lead to unemployment. There has been a lot of speculation about future of human race after the development of robots that are able to take decisions of their own. It is believed that if robots are able to take decision on their own then they may develop into a race superior to the human race. But as of now these are just ideas confined to the area of science fiction.

At the beginning of new millennium, our society has several big problems to be solved to maintain good quality of life. Those are relating to saving natural resources, increase of aged population in the society, preservation of global environment and so on. We suppose that robotics and intelligent system technologies can be a powerful way and play a more important role to solve the problems of our future society will have. Actually, there are strong expectations to robotics and intelligent system technologies in our society. There are new technological trends like computer network technologies, micro-/ nano-technologies and so on, which make us expect to develop powerful robotic and intelligent system technologies applicable to new commercial markets.

Although robots have certain social issues that go against them, like every new technology, these issues are overshadowed by the benefits of using robots.

CHAPTER 11

# Computer-aided Process Planning

## 11.1 INTRODUCTION

With the rising importance of industrial competitiveness, the need for technological improvements in the manufacturing arena is becoming acute. It is clear that the source of many of these improvements will be the field of automation. Manufacturing automation can speed up product turnaround, reduce the need for retooling, and lead to a more efficient allocation of resources. Automation can be effective for small batch manufacturing and in spare parts production. While this is a desirable goal, many small shops cannot afford to fully automate. By using clearly defined interfaces, a shop can support both manual and automated operations. For fully automating, a factory should yield useful results for manual, semi-automated and fully automated facilities.

There are a number of obstacles to the implementation of a fully integrated automated manufacturing facility. One major gap is the lack of smooth information flow between *Computer-Aided Design (CAD)* systems and *Computer-Aided Manufacturing (CAM)* systems. Traditionally, these two functions have been treated as completely separate activities. There is no feedback from CAM to CAD to reflect the manufacturability of a particular design. There are very few commercial/ production systems, which actually integrate CAD with CAM.

Extending these ideas to general part families is a much more difficult task. A brief example of the information flow illustrates a number of problem areas. A client requests some product to be manufactured and provides a loose set of requirements. This request is translated into a local representation, usually a part drawing. This local representation includes some simple translation of functional attributes into specific tolerance information. The information is loosely organized as notes or text on the part drawing. This step can be done manually or through CAD. The step is complete when the client and designer agree that the drawing adequately represents the client's needs. The problem with this approach is that the information is not represented in a computer database form. This implies that a human will have to interpret the notes sometime later in the process, leading to ambiguities. More importantly, this approach does not allow any feedback to the designer as to the manufacturability of the design. The next step in the process is to bridge the link to the CAM systems. This is called process planning. Process planning transforms the design information into some local process specification structure used by the manufacturing organization. This step includes defining a group of machinable features and their associated processing steps, selecting target machine tools to be used to process the part, generating tool and fixturing orders, and any other information needed to actually produce the part. The CAM system then expands each process step into more detailed instructions including robot or NC machine tool programmes, tool offsets, etc. It is at this point that important information is generated which should be communicated back to the designer. The important point is to produce a product at minimum cost while retaining the desired quality and functionality. The above described information flow is illustrated in Figure 11.1.

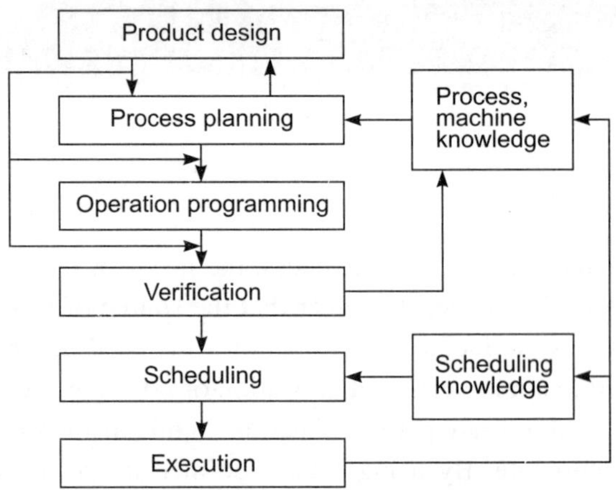


**Figure 11.1** Product realization

Thus, the step called process planning is the transformation of information from the CAD representation to the CAM representation. The transformation rules that engineers apply are not well understood even by those who use them. Clearly this makes it difficult to encode those rules in process planning systems. It is only when these rules can be represented in automatic systems that any feedback can be given during the design process. To accomplish this, a more powerful product representation is needed. This representation must serve the needs of the designer who is striving for functionality, as well as the manufacturing engineer who wants high quality at low cost. As many computer-aided tools support the design process, computer-aided process planning (CAPP) has evolved to simplify and improve process planning and achieve more effective use of manufacturing resources. In essence, CAPP can be considered as a bridge between CAD and CAM.

## 11.2 METHODOLOGY IN PROCESS PLANNING

Because of the interrelationship between the product design, material selection and process planning, some manufacturing engineering decisions need to come early in the design process, even while the design is in the conceptual stage.

In order to prepare a process plan, a process planner needs to acquire the following knowledge:

- Ability to interpret an engineering drawing
- Familiarity with manufacturing processes and practice

- Familiarity with tooling and fixtures
- Knowledge about the resources available in the shop:
  - Knowledge how to use reference books, such as machinability data handbook
  - Ability to do computations on machining times and cost
  - Familiarity with raw material
- Knowledge of the relative costs of processes, tooling and raw materials.

To prepare a process plan, the following steps need to be taken:

- Study the overall shape of the part. Use this information to classify the part and determine the type of manufacturing workstation needed.
- Thoroughly study the drawing. Try to identify all manufacturing features and notes.
- If raw stock is not given, determine the best raw material shape to use.
- Identify datum surfaces. Use information on datum surfaces to determine the set ups.
- Select machines for each set up.
- For each setup, determine the rough sequence of operations necessary to create all the features.
- Sequence the operations determined in the previous step. Check whether there is any interference between operations or dependency between operations. Use this information to modify the sequence of operations.
- Select tools for each operation. Try to use the same tool for several operations if it is possible. Keep in mind the trade off on tool change time and estimated machining time.
- Select or design fixtures for each set up.
- Evaluate the plan generated thus far and make necessary modifications.
- Select cutting parameters for each operation.
- Prepare the final process plan document.

Process planning encompasses the above mentioned activities and functions to prepare a detailed set of plans and instructions to produce a part. The planning begins with engineering drawings, specifications, parts or material lists and a forecast of demand. The results of the planning are:

- Routings, Route sheet which specify operations, operation sequences, work centers, standards, tooling and fixtures. This routing becomes a major input to the manufacturing resource planning system to define operations for production activity control purposes and define required resources for capacity requirements planning purposes.
- Process plans, which typically provide more detailed, step-by-step work instructions including dimensions, related to individual operations, machining parameters, set up instructions, and quality assurance checkpoints.
- Fabrication and assembly drawings to support manufacture (as opposed to engineering drawings to define the part).

Manual process planning is based on a manufacturing engineer's experience and knowledge of production facilities, equipment, their capabilities, processes, and tooling. Process planning is very time consuming and the results vary based on the person doing the planning. This methodology of process planning is aptly illustrated in Figure 11.2.

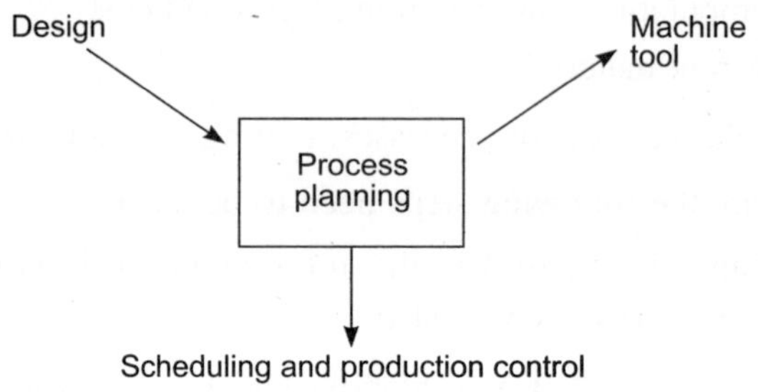


**Figure 11.2** Process planning

## 11.3 EVOLUTION OF CAPP

Manufacturers have been pursuing an evolutionary path to improve and computerize process planning in the following five stages:

Stage I - Manual classification; standardized process plans

Stage II - Computer maintained process plans

Stage III - Variant CAPP

Stage IV - Generative CAPP

Stage V - Dynamic, generative CAPP.

Prior to CAPP, manufacturers attempted to overcome the problems of manual process planning by basic classification of parts into families and developing somewhat standardized process plans for parts families (Stage I). When a new part was introduced, the process plan for that family would be manually retrieved, marked-up and retyped. While this improved productivity, it did not improve the quality of the planning of processes and it neither took into account the differences between parts in a family nor improvements in production processes.

Computer-aided process planning initially evolved as a means to electronically store a process plan once it was created, retrieve it, modify it for a new part and print the plan (Stage II). Other capabilities of this stage are table-driven cost and standard estimating systems.

This initial computer-aided approach evolved into what is now known as "variant" CAPP. However, variant CAPP is based on a *Group Technology (GT)* coding and classification approach to identify a larger number of part attributes or parameters. These attributes allow the system to select a baseline process plan for the part family and accomplish about ninety percent of the planning work. The planner will add the remaining ten percent of the effort for modifying or fine-tuning the process plan. The baseline process plans stored in the computer are manually entered using a broader concept, that is, developing standardized plans based on the accumulated experience and knowledge of multiple planners and manufacturing engineers (Stage III).

The next stage of evolution is toward generative CAPP (Stage IV). At this stage, process planning decision rules are built into the system. These decision rules will operate based on a part's group technology or features technology coding to produce a process plan that will require minimal manual interaction and modification (e.g., entry of dimensions).

While CAPP systems are moving more and more towards being generative, a pure generative system that can produce a complete process plan from part classification and other design data is a goal of the future. This type of purely generative system (in Stage V) will involve the use of artificial intelligence type capabilities to produce process plans as well as fully integrated in a CIM environment. A further step in this stage is dynamic, generative CAPP, which would consider plant and machine capacities, tooling availability, work center and equipment loads, and equipment status (e.g., maintenance downtime) in developing process plans.

The process plan developed with a CAPP system at Stage V would vary over time depending on the resources and workload in the factory. For example, if a primary work center for an operation(s) was overloaded, the generative planning process would evaluate work to be released involving that work center, alternate processes and the related routings. The decision rules would result in process plans that would reduce the overloading on the primary work center by using an alternate routing that would have the least cost impact. Since finite scheduling systems are still in their infancy, this additional dimension to production scheduling is still a long way off.

Dynamic, generative CAPP also implies the need for online display of the process plan on a work order oriented basis to ensure that the appropriate process plan was provided to the floor. Tight integration with a manufacturing resource planning system is needed to track shop floor status and load data and assess alternate routings vis-à-vis the schedule. Finally, this stage of CAPP would directly feed shop floor equipment controllers or, in a less automated environment, display assembly drawings online in conjunction with process plans.

The development of CAPP through the years is illustrated in Figure 11.3.

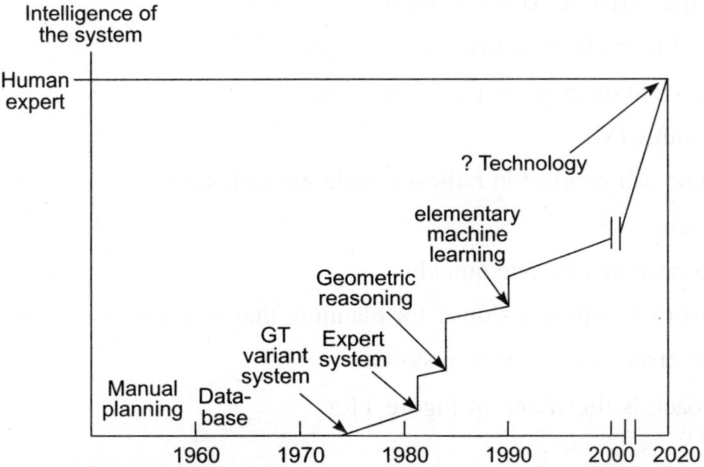


**Figure 11.3** Evolution of CAPP systems

## 11.4 TYPES OF CAPP SYSTEMS

There are two main types of CAPP systems as stated in the previous section:

(i) Variant type

(ii) Generative type.

A number of CAPP systems combine both methods, so that a third category is now recognized, the semi-generative method. It is an interim method to the time of complex development generative method.

### 11.4.1 Variant CAPP

The variant approach to CAPP was the first approach used to computerize process planning. Variant CAPP is based on the concept that similar parts have similar process plans. The computer is used as a tool to assist in identifying similar process plans, as well as in retrieving and editing the plans to suit the requirements for specific parts. Variant CAPP is related to part classification and GT coding. In these approaches, parts are classified and coded based upon several characteristics or attributes. A GT code can be used for the retrieval of process plans for similar parts. The initial challenge in developing the GT classification and coding structure for the part families and in manually developing a standard baseline process plan for each part family.

The basic variant approach to process planning is as follows:

- Go through normal GT setup procedures.
- After part families have been identified, develop standard process plans for each.
- When a new product has been designed, get a GT code for each part.
- Use the GT system to look up which part family is the closest match, and retrieve the standard plan for that part family.
- Edit the standard plan so that values now match the new design parameters, and add or delete steps as required.

Some benefits of the variant CAPP system are:

- It is well suited to medium to low product mixes
- It can be developed quickly for most companies
- Can be used with CIM
- One programme can be used in radically different industries

Its disadvantages are:

- GT codes can become obsolete quickly
- While it is fast to set up, it is slower for planning than generative systems
- More prone to error than generative systems

This variant approach is illustrated in Figure 11.4.

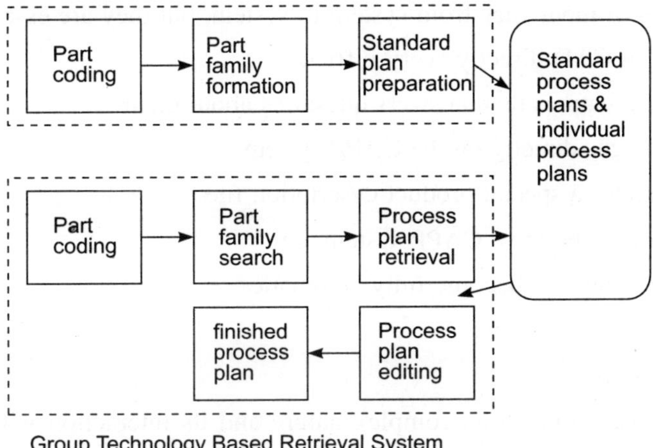


**Figure 11.4** Architecture of variant CAPP system

### 11.4.2 Generative CAPP

Generative CAPP has come into development in the late seventies. It aims at the automatic generation of process plans, starting from scratch for every new workpiece description. Often, the workpiece description is a CAD solid model, as this is an unambiguous product model. The model will contain not only shape and dimensioning information, but also tolerances, which are special features. The process plan can be routed directly to the production planning system and production control system. Time estimates and resource requirements can be sent to the production planning system for scheduling. The NC part programme and material handling control programme can also be sent to the control system. A manufacturing database, decision making logic and algorithms are the main ingredients of a generative CAPP system. In the early eighties, knowledge based CAPP made its introduction using AI techniques.

The first key to implementing a generative CAPP system is the development of decision rules appropriate for the items to be processed. These decision rules are specified using decision trees, computer languages involving logical "if-then" type statements, or artificial intelligence approaches with object oriented programming.

The nature of the parts will affect the complexity of the decision rules for generative planning and ultimately the degree of success in implementing the generative CAPP system. The majority of generative CAPP systems implemented to date have focused on process planning for fabrication of sheet metal parts and less complex machined parts. In addition, there has been significant recent effort with generative process planning for assembly operations, including PCB assembly.

A second key to generative process planning is the available data related to the part to drive the planning. Simple forms of generative planning systems may be driven by GT codes. GT or *Features Technology (FT)* type classification without a numeric code may be used to drive CAPP. This approach would involve a user responding to a series of questions about a part that in essence capture the same information as in a GT or FT code. Eventually, when features-oriented data is captured in a CAD system during the design process, this data can directly drive CAPP.

Possible sources of input vary from system to system, but they are essentially:

- Designs from CAD directly (very difficult)
- User defined features then answers questions about them
- The user designs directly on the CAPP system
- The user creates a special product description file

The advantages of generative CAPP system are:

- It has the potential to become fully automatic
- Runs faster when planning.

Its disadvantages are:

- It is still elusive due to its complex nature and its interaction with the vast manufacturing information. Only for simple shapes like axi-symmetric and prismatic parts, considerable success has been reported.
- Heavy computing resources are required.

Some of the approaches for generative CAPP are:

- Decision trees
- Decision tables
- Axiomatic approach
- Artificial intelligence-based approach and expert systems.

### Decision Trees

A decision tree is comprised of a root and a set of branches originating from the root. In this way paths between alternate courses of action are established. Branches are connected to each other by nodes, which contain a logic operation such as an "and" or "or" statement. When a branch is true, travelling along the branch is allowed until the next node is reached, where another operation is assigned or an action is executed. Decision trees can call one another recursively. Decision trees can either be used as computer or represented as data. As a computer code, the tree is converted to a flowchart. The starting node is root and every branch represents a decision statement, which is either false or true. Decision trees have certain definite benefits over decision tables:

- Trees can be updated and maintained more easily than can decision tables
- Selected branches of the decision tree may be extended to a considerable depth if necessary, while other branches may be quite short, which is more difficult to do with decision tables
- Some branches of the decision tree may be used to define TYPE, and others ATTRIBUTES, which results in relatively small trees
- Trees are easy to customize, visualize, develop and debug.

### Decision Tables

Decision tables organize conditions, actions and decision rules in tabular form. Conditions and actions are placed in rows while decision rules are identified in columns. The upper part of the table includes the conditions that must be met in order for the actions (represented in the lower part of the table) to be taken. Decision table should contain the actual rules, conditions specified in the design. According to the rule representation, decision tables can be classified as follows:

- Limited entry decision tables that represent the exact conditions (input values) as true or false entries
- Extended entry decision tables that specify the condition but not the value
- Mixed entry decision tables whereby sequenced and unsequenced actions can be entered. Sequenced actions rate a sequence number while unsequenced actions do not rate one.

### Axiomatic Approach

Its intention is to provide a logical framework for designing products and processes. A set of desired characteristics of the design, known as *Functional Requirements* (FRs) must be defined to establish a *design range*. A set of *Design Parameters* (DPs) must also be identified so that the system range can be defined. Axiomatic approach to process planning is used to define and simplify the relation between the FRs and DPs through a set of axioms. The entire framework is based on two axioms:

*Axiom 1:* The independence axiom - Maintain the independence of functional requirements,

*Axiom 2:* The information axiom - Minimize the information content.

To implement axiomatic design in process planning, the following steps are applied:

- List all the design (or production) parameters to be evaluated
- Divide the surfaces to be produced into surface groups, each of which is to be machined by a single machine
- List candidate machines for each surface group
- Evaluate all alternatives for the production and machine parameters
- Obtain the total information content and select the best machine combination based on the information content.

### Artificial Intelligence Technique and Expert Systems

Artificial intelligence techniques like formal logic, describing components and expert systems for codifying human processing knowledge are also applicable to process planning problems.

Due to the rapid development artificial intelligence techniques, generative process planning systems are built as expert systems with technological knowledge representation, which specifies general and detailed rules of process planning. They have involved object orientated programming techniques and virtual single manufacturing database techniques. Some new generation systems have employed fuzzy logic and neural network techniques and machine learning approach.

The major areas of activities in generative approach to process planning are:

- determination of part characteristic
- partial product designing
- elementary activities selection
- machine selection
- technological instrumentation selection
- technological operation sequencing
- tool selection
- machining parameters selection
- time and cost calculation.

### 11.4.3 Semi-Generative Method

*Semi-generative* method is an interim approach used when problems have occurred in building purely generative process planning system. The semi-generative method can be characterized as an advanced application of variant technology employing generative type features. Systems utilized semi-generative method to process planning should co-operate with planner, who possesses technological knowledge. The planner's responsibility is the interpretation of decision data and/or working drawing. There are following ways of this method:

- Variant method can be used to develop the general process plan, then the generative method can be used to modify it.
- Generative method can be used to create as much of the process plan as possible then the variant method can be used to fill in the details.
- Process planner can select either generative mode for complicated part features or variant mode for fast process plan generation.

## 11.5 CAD/CAM INTEGRATION AND CAPP FEATURES

A frequently overlooked step in the integration of CAD/CAM is the process planning that must occur. CAD systems generate graphically oriented data and may go so far as graphically identifying metal, to be removed during processing. In order to produce such things as NC instructions for CAM equipment, basic decisions regarding equipment to be used, tooling and operation sequence need to be made. This is the function of CAPP. Without some element of CAPP, there would not be such a thing as CAD/CAM integration. Thus CAD/CAM systems that generate tool paths and NC programmes include limited CAPP capabilities or imply a certain approach to processing.

CAD systems also provide graphically oriented data to CAPP systems to produce assembly drawings, etc. Further, this graphically oriented data can then be provided to manufacturing in the form of hardcopy drawings or work instruction displays. This type of system uses work instruction displays at factory workstations to display process plans graphically and guide employees through assembly step by step. The assembly is shown on the screen and as an employee steps through the assembly process with a switch, the components to be inserted or assembled are shown on the CRT graphically along with text instructions and warnings for each step. Some of the approaches to computer aided process planning is shown in Figure 11.5.

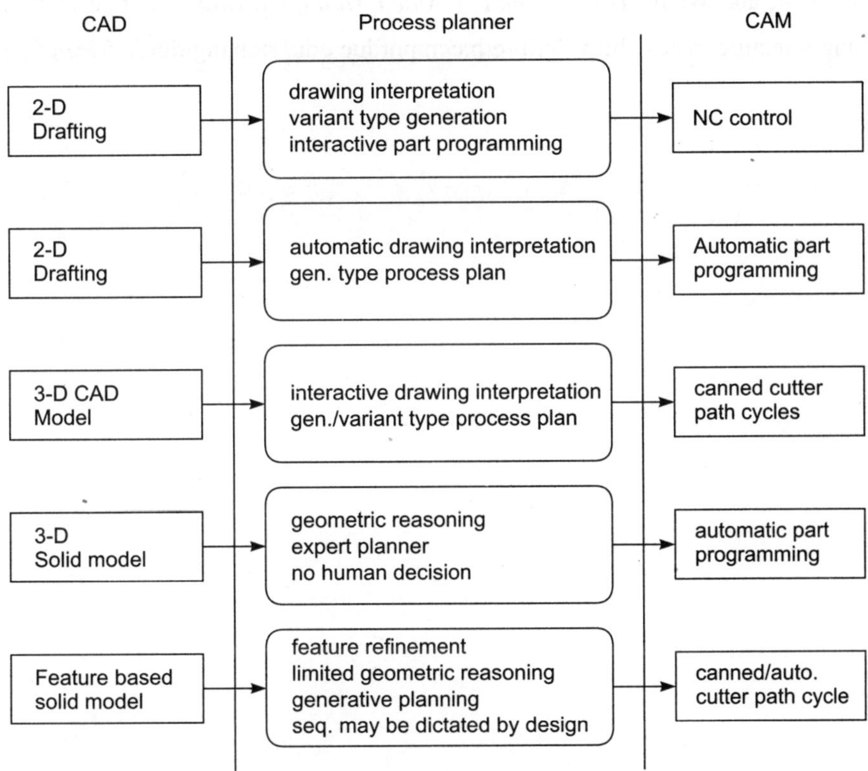


**Figure 11.5** Approaches in CAPP

## 11.6 CONCLUSION

To summarize, the advantages of computer aided process planning are typical of those accrued when any procedure is automated via computers. A brief list of these advantages are: reduced clerical effort, fewer calculations, fewer oversights in logic, immediate access to up-to-date information, consistent information, faster response to engineering or production changes, use of latest revisions, more detailed and uniform planning, more efficient use of resources.

CAPP is a highly effective technology for discrete manufacturers with a significant number of products and process steps. Rapid strides are being made to develop generative planning capabilities and incorporate CAPP into a Computer-Integrated Manufacturing (CIM) architecture.

Unfortunately, the success in completely automatic CAPP systems for realistic applications is still elusive in spite of decades of research efforts. This led to the development of alternate routes from CAD to CAM, one such being Rapid Prototyping (RP).

## REFERENCES

Zhang H. C., Alting, L., Computerized Manufacturing Process Planning Systems. Chapman & Hall, 1994.

Chang, T.C., Wysk, R.A. and Wang, H.P., *Computer Aided Manufacturing*, Prentice Hall, 1998

Link to Prof. Chang's lecture notes: http://gilbreth.ecn.purdue.edu/~tchang/doc/ie575/ie575lect.html

❑❑❑

# CHAPTER 12

# Rapid Prototyping

## 12.1 LAYERED MANUFACTURING

Introducing new products at ever increasing rates is crucial for remaining successful in a competitive global economy; decreasing product development cycle times and increasing product complexity require new ways to realize innovative ideas. In response to these challenges, industry and academia have invented a spectrum of technologies that help to develop new products and to broaden the number of product alternatives. Examples of these technologies include *Feature-Based Design*, *Design for Manufacturability* analysis, simulation, computational prototyping and virtual and physical prototyping. Most designers agree that "getting physical prototypes fast" is critical in exploring novel design concepts. The sooner designers experiment with new products, the faster they gain inspiration for further design changes. To meet this goal, during the last decade a new physical rapid prototyping concept called *Layered Manufacturing* or *Solid Freeform Fabrication (SFF)* has gained popularity world-wide. Since the aim of this family of new technologies is compression of time spent in product development and tooling, they are popularly known as *Rapid Prototyping & Tooling (RP&T).*

The principle of RP is illustrated in Figure 12.1. The CAD model of the object shown in Figure 12.1a is sliced by parallel planes. The edges of these slices thus obtained (Figure 12.1b) are squared (Figure 12.1c). Thus a complex 3D object is decomposed into several 2D objects or slices. In other words, a complex 3D manufacturing problem is converted into several simple 2D manufacturing problems. These slices are physically realized in one of several ways, stacked and pasted together as shown in Figure 12.1d, to obtain the physical prototype (Figure 12.1e). The accuracy of these prototypes, due to the staircase effect, can be improved by decreasing the slice thickness. For even better finish, polishing can be applied (Figure 12.1f).

Each physical layer will be placed over the previous one. If the previous layer is smaller than the current one, then it will not be able to fully support the current layer. For this purpose, a complementary shaped sacrificial layer of a different material is deposited and fused to the previous layer using one of several available deposition and fusion technologies. The sacrificial material has two primary roles: first, it holds the part, analogous to a "fixture" in traditional fabrication techniques; second, it serves as a substrate upon which "unconnected regions" and overhanging features can be deposited. The unconnected regions require this support since they are not joined with the main body until subsequent

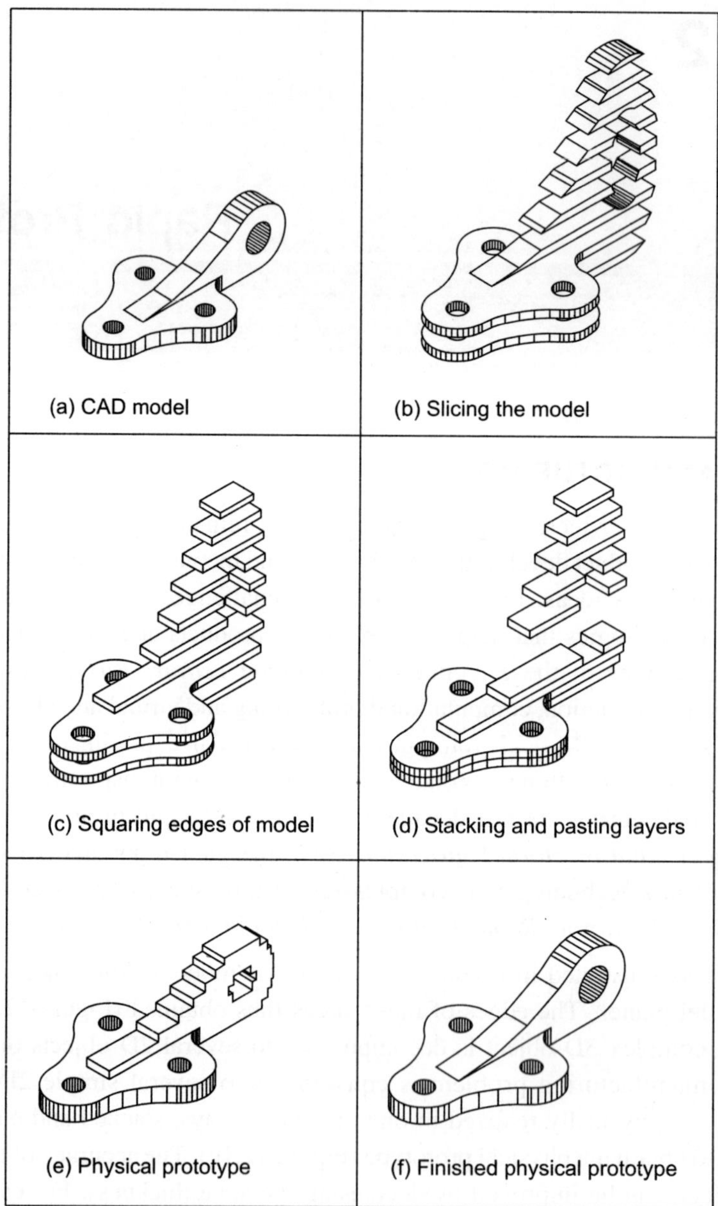


**Figure 12.1** Principle of rapid prototyping

layers are deposited. Another use of sacrificial material is to form blind cavities in the part. The collection of these sacrificial layers is called *support structure*. Features shown in Figure 12.2 require support structures.

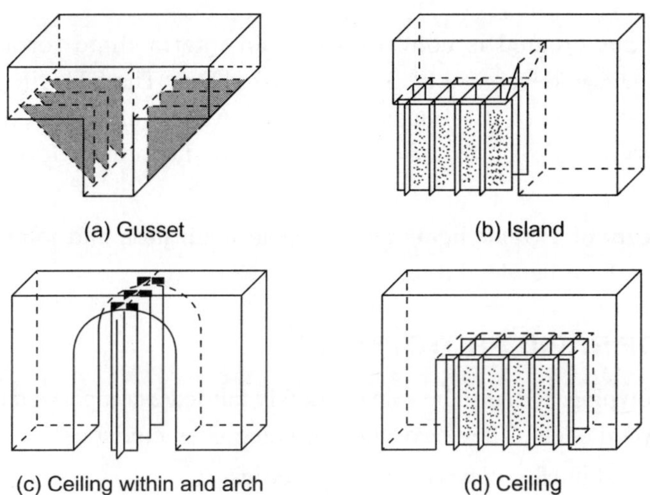


**Figure 12.2** Features requiring support

Building up structures in layers is not a new idea; it has been in practice from the days of pyramids. Any civil engineering construction follows the layer-by-layer approach. They too use support structures in the form of concrete slabs and frames when windows and doors are encountered. However, these are highly labour intensive and massive in size. In RP, complete automation from CAD to CAM has been achieved skipping steps such as process planning, tooling etc. Practical implementations of layered manufacturing for modern manufacturing needs have been made possible by several enabling technologies, including CAD-based solid modelling, lasers, ink-jet printing, and high-performance motion controllers, integrated with more traditional manufacturing processes, such as powdered metallurgy, extrusion, welding, CNC machining and lithography, into novel arrangements.

The art of building 3D objects by layers was significantly advanced by 3D Systems Inc., a U.S. company. Availability of 3D computer models was crucial to realizing the concept of layered object creation, but other technologies such as affordable laser systems, photo-curable materials, and powerful personal computers helped to disseminate this technology, called *Stereolithography*. This technology today is capable of producing highly complex 3D geometries with little or no human intervention. Emerging almost in parallel with the advancement of stereolithography were alternative systems for layered manufacturing. The layered manufacturing systems are build layered objects by lamination of sheet materials and by layered fusion or binding of powders or extruded wires (Stratasys). These processes have added a range of new materials that go beyond those of photo-curable polymers as used in stereolithography.

The RP processes differ in the way the slices are physically realized and the way they are stacked and glued together. During the 'Art to Part' conversion in RP, none of the traditional manufacturing steps such as process planning, tool design and movement of material from one machine to the other to carry out various operations take place. Thus no human intervention is required to produce the physical part.

The first step in any RP process is to create the computer model of the object either as a surface model or a solid model. Most commercially available CAD/CAM/CAE packages can be used for this purpose. The model thus created is converted into an intermediate format called STL format (STereolithography Language developed by 3D Systems) U.S.A. The STL file is the simplest form to represent an object in the form of its boundary triangles. Almost all CAD/CAM/CAE packages can output the part geometry in the STL format since it is the de-facto input format for all RP processes today.

By using this concept of virtual slicing and physical realization and joining, the family of RP& Rapid Tooling processes have heralded a new era called *Slice Age*.

## 12.2 BENEFITS OF RAPID PROTOTYPING

The goal of Rapid Prototyping (RP) is to be able to quickly fabricate complex-shaped, three-dimensional parts directly from CAD models. They have the following characteristics:

- They can build arbitrarily complex 3D geometries
- Process Planning is automatic, based on a CAD model
- They use a generic fabrication machine, i.e., do not require part-specific fixturing or tooling
- They require minimal or no human intervention to operate.

These characteristics enable quick manufacture of prototypes benefiting various groups involved in product development as described below:

*To the Product Designer*: The product designers can increase the part complexity with little effect on lead-time and cost. They can optimize part design to meet customer requirements, with little restrictions like material wastage or large thin walls, which are otherwise imposed by machining operations. There is improved design creativity and valuable feedback on the design can be obtained from various sources.

*To the Manufacturer*: The manufacturers can avail of the benefits like early realization of profit, reduction in cost due to reduction in wastage and scrap, reduced labor content, reduced inventory, assembly and inspection costs due to reduction in parts count. All testing and design changes are carried out before spending any money on tooling. Design problems can be solved before they become tooling glitches or manufacturing fiascos. The RP parts act as important communication tools between the designers, engineers and vendors. Different people interpret a 2D drawing differently and clearing this confusion costs valuable time. When everyone can look at the part itself, inspect it, even touch it - design concepts are clarified and misunderstandings disappear. Design errors like too thin walls, misaligned apertures, inner walls crossing outer walls, etc. cannot be identified in 2D drawings or 3D models. RP helps in locating and solving these problems early before costly scrapping or reworking of tools is required. The tooling can be done only when the concept is refined and the data is verified. In short, Rapid Prototyping and Tooling are manufacturing tools that enable industries to optimize information at front end of design and manufacturing process and thereby work smarter and faster.

*To the Marketing staff*: The market greatly benefits from these techniques because of the advantages like reduced time-to-market, reduced risk of product failure in the market, manufacturing of products meeting customer requirements and possibility of test-marketing of new products. Limited editions of a variety of conceptual prototypes of the same product can be launched followed by mass production of the most successful ones.

*To the Consumer*: The consumer can buy products, which meet more closely individual needs and wants. There is a much wider diversity of offerings to choose from.

However, RP parts suffer from the following severe limitations:

- There are limited material options. These materials are proprietary and hence costly.
- Finish is nowhere close to machined surfaces.
- Poor dimensional accuracy and stability.
- Anisotropy.

## 12.3 APPLICATIONS OF RP & RAPID TOOLING

Prototypes made using RP& Rapid Tooling systems find applications as:

*Concept models*: Designers always prefer to present form ideas in mock-up models or prototype of product, for final presentations. Quick RP models become very handy for this purpose. Though little expensive, the time saved is enormous and one can make very intricate details even like snap-ons to demonstrate the working of fitments.

*Models for market research*: With RP technology, several different variations in design models can be made simultaneously and the best one can be selected. Few accurate copies of the selected one could be produced in a short time to get a feel of the market.

*Rapid Tooling*: Manufactures are exploiting rapid tooling methods to make injection moulds to introduce a number of designs in the market in peak season. Final hard tooling is done only for those models toys which perform best in the market

*Tender model*: Many automotive companies have started realising the advantages of providing a physical three dimensional part model to a subcontractor for a quote. This helps the vendor to decipher the CAD file quickly and assist them in deciding the parting lines, a very important step in tool design.

*Wind tunnel models*: The models of planes, automobiles, trains, buildings and structures are tested for performance in wind tunnels. Accurate RP models can help in obtaining reasonably good results.

*Models for stress analysis*: Many RP models are directly used for experimental stress analysis of parts under loads by photostatic stress analysis and other analytical methods

*Medical applications*: RP parts are being increasingly used to manufacture one-off replicas of bones. The scanned (CT or MRI) data of affected bone areas is used to prototype the part which in turn is used to create ceramic shells for investment casting of metallic replacement parts. RP & Rapid Tooling in conjunction with Reverse Engineering will be immensely beneficial in the field of medicine.

## 12.4 IMPORTANT ISSUES IN RP

There are two very important issues in any RP process which need to be addressed:

(i) Slicing strategy and

(ii) Mechanism for support structure.

The former influences the cost and efficiency of the process for a given accuracy. The latter gives the process the ability to produce shapes of any complexity, viz. reentrant or overhanging features.

### 12.4.1 Slicing Strategies

As we discussed, in RP, the CAD model of the required 3D object is cut by several parallel planes to get the 2D slices. Each slice $S$ is associated with a pair of these planes one at the top and the other at the bottom respectively called $P_t$ and $P_b$. If the distance between $P_t$ and $P_b$, called *Slice Thickness* or *Layer Thickness* is the same for all layers of the object, it is called *Uniform Slicing*; otherwise, it is called *Adaptive Slicing*. In adaptive slicing, the local geometric properties, viz., (i) *curvature* and (ii) the orientation of the local surface with respect to the build direction are taken into account in deciding the highest possible slice thickness to meet the required accuracy. Lower the curvature, the more the surface is along the build direction and poorer the accuracy (i.e., higher the allowable deviation), higher will be the slice thickness. Higher the slice thickness lower is the time for building and hence lower is the cost of manufacture.

The geometry of each slice is in the form of pairs of contours, one contour of each pair belonging to the planes $P_t$ and $P_b$. In all popular RP systems today, the contours in any one of these two planes only are retained discarding those in the other plane. The retained contours are extruded to the other plane thus getting the approximated or squared geometry of the slice. This is called *$0^{th}$ order edge approximation* in which the edges of the layer are along the build direction. If the contours in both planes are retained and each corresponding pairs of contours are joined by a ruled surface, it is called *$1^{st}$ order edge approximation*. This is more accurate than the former as can be seen from Figure 12.3.

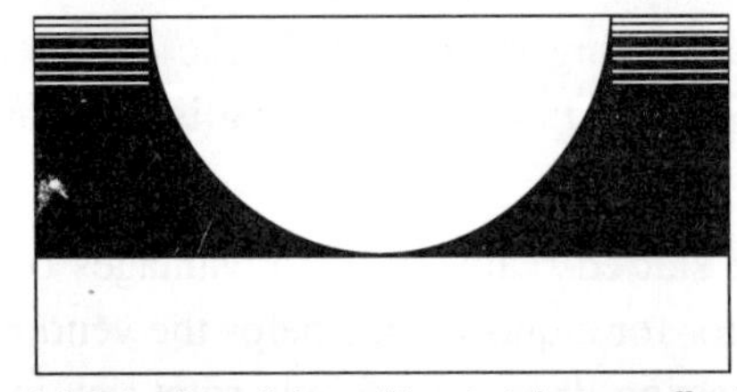

(a) Adaptive slicing without staircase effect elimination (number of layers is 251 in this case).

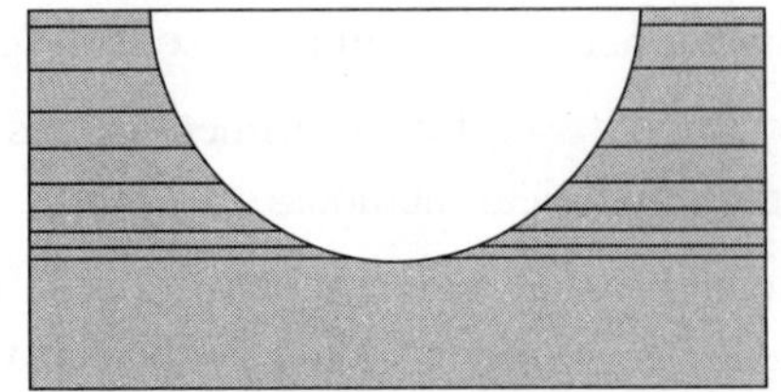

(b) Adaptive slicing with staircase effect elimination (number of layers is 11 in this case).

| | |
|---|---|
| Uniform slicing (constant layer thickness) | 750 slices |
| Adaptive slicing without staircase elimination | 251 slices |
| Adaptive slicing with staircase elimination | 11 slices |

**Figure 12.3** Comparison of various slicing strategies

In fact, one can go for even higher order approximations although beyond $2^{nd}$ order it looks impossible to realize as of now. *$2^{nd}$ order edge approximation* is possible for cutting polystyrene by appropriately flexing the cutter wherein the pair of contours at the top and bottom planes can be joined by a quadric surface.

All the existing RP processes use only *uniform slicing with $0^{th}$ order edge approximation* due to its computational simplicity and their process limitations. In order to get the same quality, it is not necessary to keep the slice thickness uniformly small throughout as shown in Figure 12.1. Adaptive slicing and higher order of edge approximation help in drastically bringing down the number of layers. Figure 12.3 illustrates the manufacture of a hemispherical cavity of 50 mm radius with a maximum deviation of 0.1 mm using the following three types of slicing:

1. Uniform slicing with $0^{th}$ order approximation
2. Adaptive slicing with $0^{th}$ order approximation and
3. Adaptive slicing with $1^{st}$ order approximation.

The number of slices in each of these methods are 750, 251 and 11 respectively. It can be seen that adaptive slicing is superior to uniform slicing and the $1^{st}$ order approximation is preferable to $0^{th}$ order approximation for the edges. Therefore, in the new process, adaptive slicing with $1^{st}$ order edge approximation is used. A RP software developed supports all four types of slicing as shown in Figure 12.4. An example of adaptive slicing of $1^{st}$ order edge approximation obtained using this software is shown in Figure 12.5. Note that it requires only 26 slices as against 440 required in uniform slicing of $0^{th}$ order edge approximation for the same accuracy.

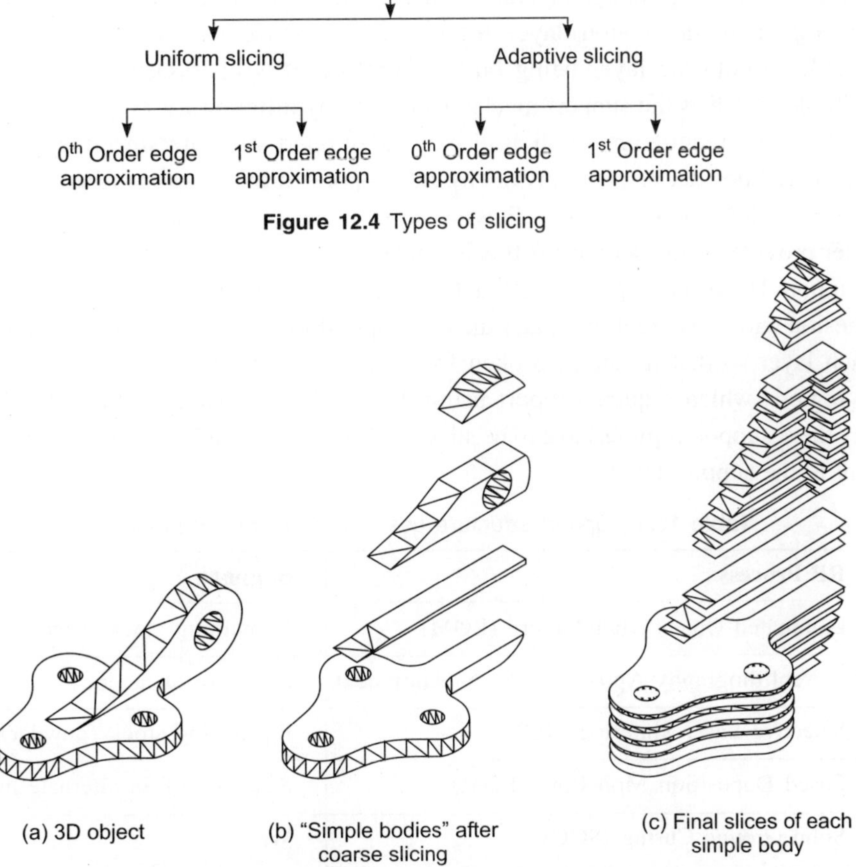


**Figure 12.4** Types of slicing

**Figure 12.5** An example of adaptive slicing of $1^{st}$ order edge approximation

Computer-Aided Process Planning or Part Programming (CAPP) is still a gray area in manufacturing that is yet to be solved for all types of shapes taking into account non-geometric constraints related to materials, dimensional and geometric tolerances, mechanics of the process etc. Reasonable CAPP solutions are available only for the manufacturing of surfaces of revolution and prismatic parts. In spite of the best efforts of several research groups, slow progress in CAPP is one of the reasons for the popularity of RP which can produce objects of any complexity without the need for any intelligent process planning. However, there is now a trend to mix as much of CAPP knowledge available with RP leading to hybrids. One such development in this direction is called *Local Adaptive Slicing* where different features at the same height will have different slicing methods depending on their orientation and geometric characteristics. For instance, two cylinders of the same dimensions but one with its axis vertical and the other with its axis horizontal will have different number of layers in this type of slicing. Another interesting development is splitting of the object with horizontal planes with each slice having simple automatically machinable features like blind holes, pockets etc. We can call these types of slicing, where layers of very high thicknesses are possible, as *Intelligent Slicing*.

### 12.4.2 Mechanism of Support Structures

When there are *re-entrant* or *undercut* features along the build direction as shown in Figure 12.2, a layer may be larger than the previous layer. If their size difference is too much, the layer at the bottom may not be able to hold the layer being built. Therefore, it is necessary to provide some support mechanism. Table 12.1 lists the support mechanisms used by different RP processes. RP processes like LOM and SLS do not require any explicit or external support structure since the remaining stock of sheet and unsintered powder respectively act as the support. In SGC, wax is filled in the gaps on each layer which acts as the support for the following layer. In SLA equipment, the buoyancy of the liquid photo-polymer provides some support; if this is not adequate, the part geometry is modified to include bristle like support. However, in processes like FDM, a separate nozzle deposits support material which has poor adhesion with the modelling material. This support material is deposited coarse at the required places in each layer so that it can be broken from the model easily. In these processes, one has to identify the regions which require support using the *concept of visibility* along the build direction. Subsequently these support regions have to be sliced and the corresponding path for the support structure has to be sent to the support head.

**Table 12.1** Support structure used for different RP processes

| RP Process | Support |
|---|---|
| Laminated Object Manufacture (LOM) | Remaining stock (sheet) |
| Stereolithography Apparatus (SLA)/equipment | Structure of model material |
| Selective Laser Sintering (SLS) | Remaining stock (powder) |
| Fused Deposition Modelling (FDM) | Structure of an alternate material |
| Solid Ground Curing (SGC) | Wax |

The material used for support structure is 'sacrificial' in nature since it is removed at the end of the building process. The removal of support structure is sometimes quite involved and may affect the surface finish on the prototype. For instance, removal of wax from the SGC RP part takes several hours and requires special vibrating arrangements and solvents. As we shall see later, building and removal of support structure is quite time consuming in LOM process.

## 12.5 POPULAR RP PROCESSES

One of the broadest ways of classifying RP systems is by the initial form of their material. On this basis, the systems can be classified as:

*Liquid-Based Processes*: Liquid-based RP systems begin with their material in a liquid state. Through a process commonly known as curing, the liquid is converted into a solid state. The important processes under this category are SLA and SGC.

*Solid-Based Processes*: The material in these processes can be either in the form of a wire, a roll, laminates and pellets. The important processes under this category are LOM and FDM.

*Powder-Based Processes*: These processes use powder in grain-like form. Principal amongst these processes are SLS and 3DP.

Some of the most popular RP processes are:

- Laminated Object Manufacturing (LOM)
- Fused Deposition Modelling (FDM)
- Stereolithography Apparatus (SLA) equipment
- Solid Ground Curing (SGC)
- Selective Laser Sintering (SLS)
- Three-Dimensional Printing (3DP)

These processes are discussed in the following sections.

### 12.5.1 Laminated Object Manufacturing (LOM)

*Laminated Object Manufacturing (LOM)* is a RP process that was developed and commercialized by Helisys Corporation, USA. LOM builds shapes with layers of paper or plastic. The sheet is available in the form of a roll with a thermally activated adhesive on one side. Every time, the paper roll indexes by a constant distance over a table where building of the part is taking place. Using a heated roller, this new layer or laminate is glued to the previous layer. A laser cuts the outline of the part cross-section for each layer. The strength and focussing of the laser is such that the depth of cut is just equal to the laminate thickness. The schematic diagram of the process is shown in Figure 12.6. The laser then scribes the remaining material in each layer into a cross-hatch pattern of small squares, and as the process repeats, the cross-hatches build up into tiles of support structure. The cross-hatching facilitates removal of this tiled structure when the part is completed. The laser spends hardly 5% of the time in cutting the contours of the part while the remaining 95% of the time is wasted in cutting the stock. The removal of stock called 'decubing' to extract the RP part after building by LOM is shown in Figure 12.7.

Depending on the complexity of the shape, decubing may take several hours. A better method of optimal stock removal for LOM will take less time for cutting and stock removal.

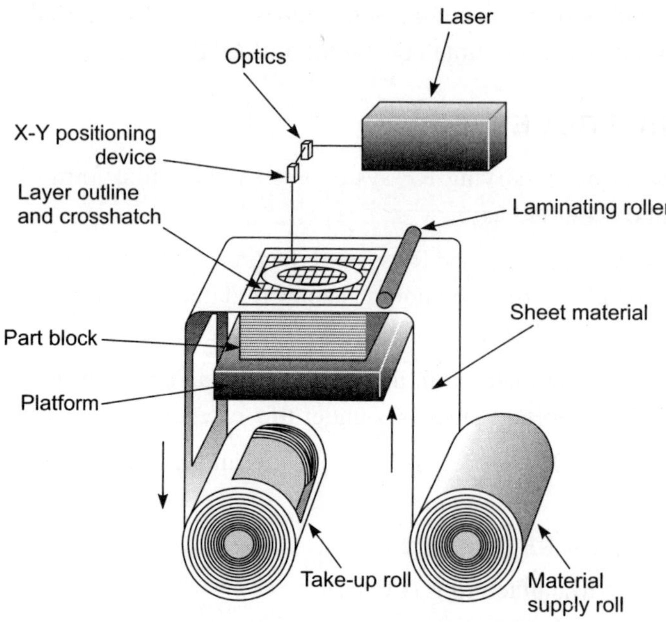


**Figure 12.6** Laminated Object Manufacturing (LOM)

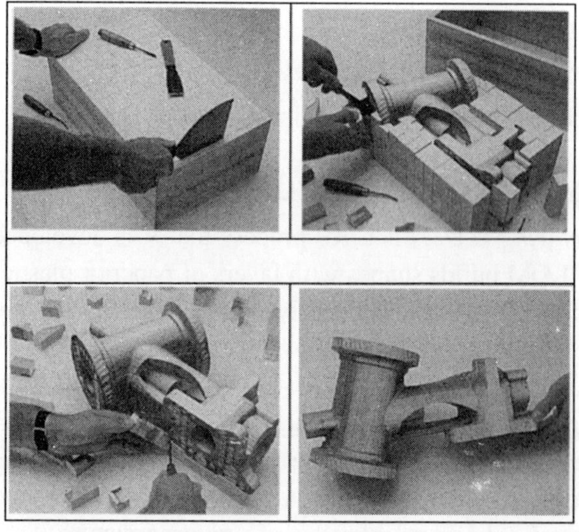

**Figure 12.7** Illustration of decubing in LOM

LOM builds up large parts relatively rapid because only contours are scanned. Internal cavities are hard to form with LOM, since it is difficult to remove the sacrificial material from the internal regions.

**Advantages**

- Only the outline is cut and no time is spent in building the interior of the layer. Therefore, this process is fairly faster.
- The materials used for building the parts (viz., wood and paper) are the least expensive among all RP processes.
- Cost of the machine is one of the lowest.
- No external support structures or post-curing is required.
- It is suitable and economical for making large parts to be used as patterns for sand castings.
- The process can be carried out unattended.
- LOM is also a direct Rapid Tooling process. It has been successfully used in making metallic laminated tools for sheet metal forming operations.

**Limitations**

- Parts are weak along Z-direction.
- Paper parts have poor surface finish and absorb moisture.
- The process is not suitable for making small intricate parts. As the stock needs to be chipped out during decubing, it requires a fair amount of skill is required.
- There is a lot of material wastage.

### 12.5.2 Fused Deposition Modelling (FDM)

Fused Deposition Modelling (FDM) was first developed and commercialized by Stratasys, Inc., USA. In this approach, a continuous filament of a thermoplastic material (polymer or wax) through a resistively heated nozzle is deposited to fill the contours of the desired slice. An explanatory sketch of the FDM process is shown in Figure 12.8. The raw material is in the form of a wire of about 3 mm diameter.

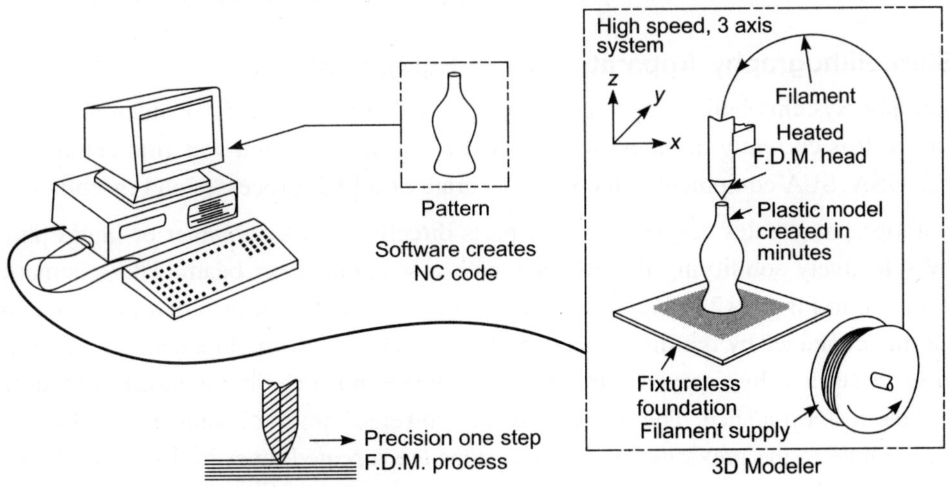


**Figure 12.8** Fused Deposition Modelling (FDM)

Using a pinch wheel drive, it is fed into an extrusion chamber which is kept at a temperature slightly above its flow point. The thermoplastic wire itself acts as the piston initially in the extrusion chamber which subsequently gets melted and pushed out through the nozzle. The filament coming out of the nozzle solidifies relatively quick after it exits the nozzle. It may be noted that the diameter or width of the filament deposited need not be same as that of the nozzle since it also depends on the ratio of the wire speed into the extrusion chamber and the traversal speed of the nozzle. It is possible to form short overhanging features without the need for explicit support in this process. In general, however, explicit supports are needed. These support structures are drawn out as thin coarse wall sections that can easily be removed upon completion. There is a separate extrusion head for depositing support material.

**Advantages**

- The process is very simple and the machine is less expensive.
- A variety of materials can be used and the material changeover, which involves only changing the head, is very fast and simple.
- No post-curing is required.
- There is little wastage of material.
- The part building can be carried out unattended.
- The wire material has a large shelf life and remains unaffected if not removed from the packing provided.

**Limitations**

- Surface finish and delicate features are inferior to other processes.
- The process is slow since the entire contour is to be filled.
- The strength is low in the vertical direction.
- Accuracy and surface finish is poorer as compared to the other RP processes.

### 12.5.3 Stereolithography Apparatus (SLA) Equipment

Stereolithography systems build shapes using light to selectively solidify photocurable resins called photopolymers. It is currently the most widely used RP technology and was first commercialized by 3D Systems, USA. SLA equipment is the most accurate of all RP processes and machines till date.

Stereolithography creates acrylic or epoxy parts directly from a container of liquid photocurable polymer by selectively solidifying the polymer with a scanning laser beam. The schematic process diagram is shown in Figure 12.9. Parts are built up on an elevator platform that incrementally lowers the part into the container by the distance of the layer thickness. To build each layer, a laser beam is guided across the surface, by a servo controlled galvanometer mirrors, drawing a cross-sectional pattern in the XY plane to form a slice. The platform is then lowered into the container and the next layer is drawn which adhere to the previous layer. These steps are repeated layer-by-layer, until the complete part is built up. Since the photopolymers are relatively viscous, simply lowering the elevator by the small distance of the layer thickness (of the order of 0.050 mm to 0.50 mm) down into the tank (container) does not permit the liquid to uniformly recoat the upper surface of the part in a timely fashion. Therefore, a recoating mechanism is required to facilitate this process. Stereolithography uses a "deep dipping" recoating, wherein the elevator is first lowered several millimeters so that the liquid entirely flows over the current upper surface of the part. The elevator is then raised to the desired height and a "blade" (wiper arm) traverses the surface to quickly level the excess viscous material. In the SLA machine, complete polymerization does not take place. Therefore, after all layers are built, the excess liquid polymer is drained and the prototype is put in a separate chamber with a flood of light to complete the polymerization.

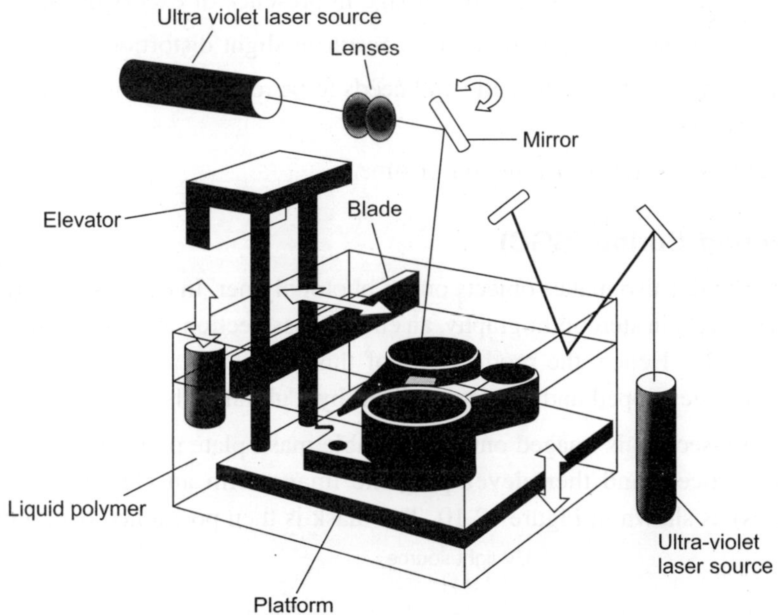


**Figure 12.9** Stereolithography Apparatus (SLA)

Features with gradually changing overhangs can be built up without support structures. The buoyancy of the viscous raw material supports the layer to some extent. However, large overhanging features require supports since the initial thin layers that form them can warp or break off as the part moves down into the liquid. The supports are typically built up as thin wall sections or bristles-like structures that can easily be broken or cut from the part upon completion.

QuickCast is a patented SLA process from 3D Systems, U.S.A. where the part is made hollow with an interior honey comb structure. This is used as a consumable pattern in shell casting.

**Advantages**

- Accuracy of ±0.050 mm and surface finish obtained are the best amongst all the processes.
- Model building can take place unattended.
- Capable of high detail and thin walls.
- SLA is also used in direct Rapid Tooling.

**Limitations**

- Experience and expertise is required in deciding support structures. The model has to be modified for this purpose.
- Material is toxic and hazardous.
- Part strength is less and may undergo warpage in presence of excess moisture.
- Post-curing of the part is required and may result in slight distortion.
- The raw material has a finite shelf-life and needs to be replaced (even if unused) after a period of about two years.
- The part becomes brittle over a period of time.

## 12.5.4 Solid Ground Curing (SGC)

Solid Ground Curing (SGC) also makes objects out of photopolymer. In contrast to "drawing out" each cross-section by a laser light in stereolithography, an entire cross-section in a single operation is exposed using photomasks in SGC. Hence, the productivity of this process is many times that of SLA. This approach was originally developed and commercialized by Cubital of Israel.

In SGC, each cross-section is imaged onto an erasable mask plate produced by charging the plate via an iono-graphic process and then developing the image with an electrostatic toner (like the photocopying process) as shown in Figure 12.10. The mask is then positioned over a uniform layer of

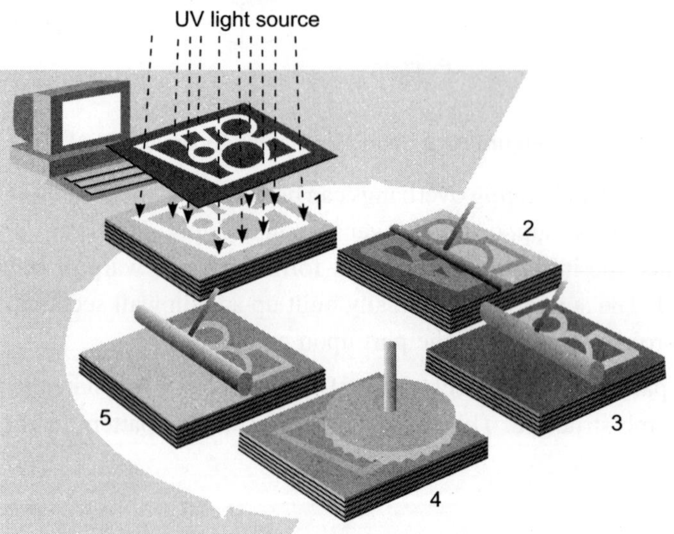


**Figure 12.10** Solid Ground Curing (SGC)

liquid photopolymer, and an intense pulse of UV light is passed through it to cure the material in the unmasked zones. Uncured photopolymer is removed from the layer with a vacuum system and replaced with a low melting point, water soluble wax that serves as the sacrificial support. After the wax has cooled, the layer is milled to produce a flat surface. As can be seen from the figure, this process involves two cycle, one for the preparation of the mask and the other for the preparation of the physical model and support parts of the layer.

The pattern on the exposed mask is erased by wiping off the toner, and the entire process is repeated. After the part has been completed, the wax is removed by melting. The various processes used to implement SCG are performed at different stations. A unique feature of the photomasking approach is its capability to build multiple parts in a single batch by packing them within the working volume.

### Advantages

- This is one of the fastest processes since curing takes place simultaneously at all desired points.
- Its high productivity owing to its ability to built a set of nested parts make it well suited for prototyping. Build time is independent of the number of parts being made at a time. Therefore, it can act as a production machine also.
- No external support structures are required.
- No warping or curling of the part takes place, as there is no post-curing operation.
- Large variety of photopolymers can be used for building the parts.
- Accuracy of parts is good.

### Limitations

- There is lot of wastage of material and wax. The resin picked up by the aerodynamic wiper and vacuum during the milling process cannot be reused. Additionally, the material which does not form the part of the model, but gets exposed to the UV light needs to be replaced. Only fifty percent of this affected material can be converted into usable form.
- The cost of the machine is the highest amongst all RP machines since involves several systems such as photo-masking system, vacuum system, milling system etc.
- The process operation is complex and maintenance cost is high since it has several subsystems.
- It requires a huge compressor and its operation is noisy.
- Monitoring of the building process is required.
- Raw material has a finite shelf life and needs to be replaced after a certain period, even if not used.
- Wax is sticky and difficult to remove.

### 12.5.5 Selective Laser Sintering (SLS)

Selective Laser Sintering (SLS) process was originally developed at the University of Texas at Austin in USA and then commercialized by DTM Corporation, USA. In SLS, a layer of powdered material is spread out and levelled over the top surface of the growing structure (Figure 12.11). A laser then selectively scans the layer to fuse those areas defined by the geometry of the cross-section; the laser energy also fuses layers together. The unfused material remains in place as the support structure. After each layer is deposited, the platform lowers the part by the thickness of the layer, and the next layer of powder is deposited. When the shape is completely built up, the part is separated from the loose supporting powder. Several types of materials are in use, including plastics, waxes, and low melting temperature metal alloys.

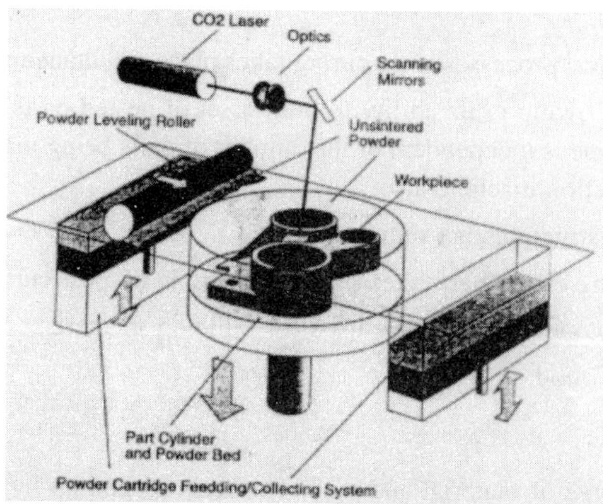


**Figure 12.11** Selective Laser Sintering (SLS)

This process has been successfully proved for making steel die inserts for short run production. For making steel dies on DTM's SLS machine, the raw material is steel powder with each steel particle coated with a polymer that acts as binder. The same machine is used for nonmetals as well as steel prototypes. When the building takes place, only the binder coating is fused keeping the particles together. Thus what is obtained at the end is a green part. This green part is put in a special oven to complete the sintering when the binder evaporates leaving it a porous part. Subsequently, it is put inside another chamber for several hours to impregnate the pores with copper. Copper impregnation is required both to get dense parts as well as good polishability. In SLS process, there is one machine for each material, viz., EOSINT-P for polymer, EOSINT-C for ceramic and EOSINT-M for metallic prototypes. There is no binder coating on the metallic particle and the metallic powder is not strong steel but one with lower melting point. The laser used for making metallic parts is sufficiently powerful to fuse the metallic particles. The metallic particles in this process apparently do not require post-sintering as well as copper impregnation. However, its laser is more powerful.

For making ceramic moulds, the sand particles are coated with a binder as is done for steel tools in the case of DTM's SLS process.

### Advantages

- Any material that can be converted into powders and can be bonded together by fusing its particles at a reasonably low temperature (about 350–500°C) can be used for making the parts in SLS process. Materials commonly used for making parts in this process are nylon, ABS and Investment Casting Wax (ICW).
- This is the only commercially available direct RP process to make prototypes out of metals. Hence, this is useful for tool makers. (For making dies and moulds).
- This can also produce ceramic mould cavities directly and hence there is no need for patterns.
- Parts obtained are tough.
- No external support structures are required.
- No post curing is required for nonmetals. Only metal parts require sintering.
- Functional metal and ceramic parts can be obtained.
- There is no wastage of material.

### Limitations

- This is one of the costliest processes.
- Surface finish of parts is grainy.
- Parts are porous in nature.
- The building operation needs to be monitored.
- Long time is required to heat up the material chamber before building the parts and to cool it down after the building is over.
- The parts are brittle.

## 12.5.6 Three-Dimensional Printing (3DP)

The first process that successfully demonstrated "printing" of shapes was Three-Dimensional Printing (3DP) process developed at MIT, U.S.A. as a method to form "green" preforms for Powder Metallurgy (PM) applications.

3DP shown in Figure 12.12 is very similar in principle to SLS with the laser beam replaced by a nozzle of binder. In 3DP, the part is built up in a bin that is fitted with a piston to incrementally lower the part into the bin. Powder (such as alumina) is dispensed from a hopper above the bin, and a roller is used to spread and level the powder. An inkjet printing head scans the powder surface and selectively injects a binder (such as colloidal silica) into the powder. The binder joins the powder together into those areas defined by the geometry of the cross-section. The unbound powder becomes the support material. This results in a green part as in the case of SLS and hence subsequent post-processing is also similar to SLS. When the shape is completely built up, the "green" structure is fired, and then the part is removed from the unbound powder. 3DP of metal powders, such as stainless steel bound with a polymeric binder, is also being explored; subsequent infiltration of the matrix is then required for densification.

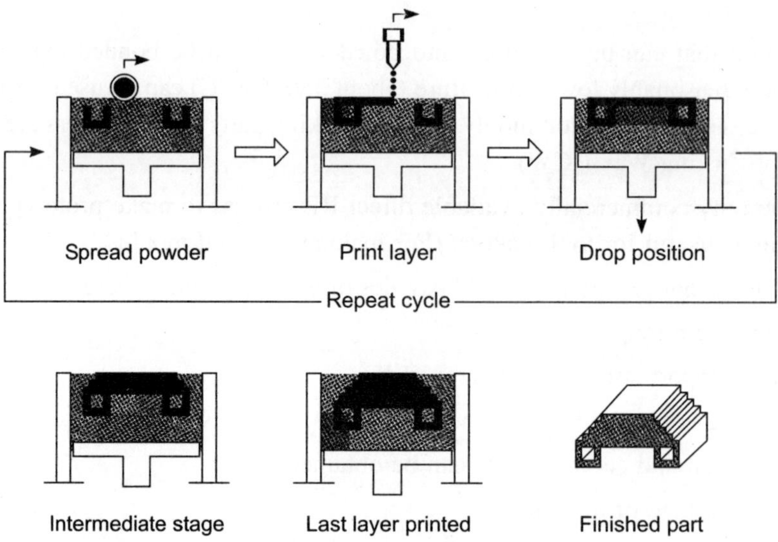


**Figure 12.12** 3D Printing (3DP)

Companies are making RP machines using this principle for making models out of starch, sand and metal.

### Advantages

- Functional metal parts can be obtained directly. This is its most important advantage.
- Ceramic shells for investment casting are obtained directly thus eliminating tooling required for making wax patterns.
- No external support structures are required.

### Limitations

- Surface finish of parts is poor.
- Parts are likely to be porous.
- Accuracy is inferior to other processes.
- Post curing is required.

## 12.6 CONCLUSIONS

RP can produce any complex geometry from its CAD model automatically without the need for any tooling. Although it was initially introduced for shape visualization, it is now being used for limited production, functional testing and tool making both in direct and indirect routes. This technology is becoming more and more popular due to slow progress in CAPP which is a link or interface between

CAD and CAM. A trend is seen today where people are trying to absorb the accomplished features of CAPP and RP technologies so that higher slice thickness and higher accuracy are possible at the lowest cost.

STL format continues to be the standard mode of geometry transfer between CAD systems and RP machines due to its simplicity. However, these files are too huge. This is also error-prone since the topological information to define the solid is not explicitly available. However, so far no alternate format to STL has emerged in spite of tremendous research efforts. Therefore, it is expected that STL format will continue for quite some time and hence special packages to browse STL files and repair them are available today. Since some areas such as RP and VR make use of polyhedral geometry, polyhedral geometric medelling kernels will be developed in future that will help rapid software development in these areas.

# CHAPTER 13

# Rapid Tooling

## 13.1 INTRODUCTION

Moulds, patterns and die cavities are the commonly used tooling elements. The following are the two common types of patterns:

(i) *Permanent patterns*: These are popularly used for sand casting. Depending on the quantity and quality requirements, these could be made of wood or metal.

(ii) *Consumable patterns*: When the pattern gets destroyed during the process of casting, it is called consumable pattern. Hence, a production line for making these patterns is required. Sand cores also can be considered in this category. Wax patterns are used in *Investment Casting (IC)* and polystyrene or thermocole patterns are used in *Lost Foam Casting (LFC),* also known as *Evaporative Pattern Casting (EPC)* or *Full Mould Casting (FMC).*

Similarly, the die cavities or moulds also can be of two types:

(i) *Permanent moulds*: When the die halves can be used to make several parts before they wear off, they are called permanent moulds. Processes like *Injection Moulding* (for making plastic parts) and *Die Casting* (for making light metal parts) use this type. Generally, these are made of die steel. When the quantity is less, they can also be made of brass or other lighter materials.

(ii) *Consumable moulds*: When the mould gets destroyed during the process of casting, it is called consumable mould. Common example for this is the mould of sand casting.

RP introduced in 1989 as a design visualization tool has revolutionized the way products are designed and manufactured today. It helped in compressing the product development time tremendously. However, the high cost of RP parts cannot be justified if its use is restricted to visualization alone. Therefore, need was felt to use this technology for making one or more of the above tools. When RP technology is extended to make the tooling elements such as moulds and dies, it is called *Rapid Tooling* or *Rapid Prototyping & Tooling (RP&T or RPT)*. These RT processes can be classified into the following (Figure 13.1):

(i) Direct processes

(ii) Indirect processes.

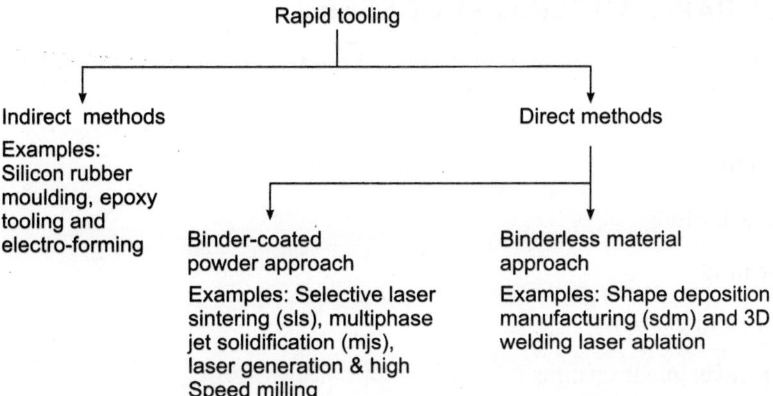


**Figure 13.1** Classification of rapid tooling processes

If the tooling element is made directly in a layer by layer manner, such a process is called direct RT process. On the other hand, when a RP part made of a material other than the tooling material, and the geometry of this prototype is transferred to the actual tooling material, such a process will be referred as indirect process. Indirect RT processes came first, to be followed by direct RT processes.

Although the prototypes could be made out of only soft materials, very soon, innovative methods of using them for making the short run and production tools were developed using the existing technologies such as *Silicon Rubber Moulding*, *Epoxy Tooling* etc. These are indirect RT processes. As these processes require a master model that can be easily and accurately produced, these processes gained further importance.

Processes such as SLS can produce metallic and ceramic tools and hence are called Direct RT Processes. In these processes, the raw material is the required hard material coated with a soft binder. While selectively sintering during the layer building process, only the soft material melts binding the hard particles around it. The prototype thus obtained is initially in 'green state' that needs to be post-sintered and/or infiltrated with a low melting material to get the desired mechanical properties for the tool. The properties of these tools will be invariably inferior to the conventional tools since the hard particles do not fuse together fully and the density is lower due to the presence of voids. These processes also suffer from shrinkage related inaccuracies. Such processes following a "binder coated powder approach" are more suitable for only short runs. LOM process has been successfully used to produce tools for sheet metal as well as injection moulded parts. However, these laminated tools suffer from poor bonding between the layers due to mechanical fastening and their inherent inability to have adaptive layer thickness.

As the direct processes are more accurate and desirable, a lot of research works are going on in this area. These details of the direct and indirect processes will be discussed in this chapter.

## 13.2 INDIRECT RAPID TOOLING PROCESSES

The following indirect RT processes will be discussed here:

- Silicone rubber moulding
- Epoxy moulding
- Spray metal tooling
- Electroforming
- Pattern for sand casting
- Pattern for investment casting
- Polyurethane tooling
- Vacuum casting
- Vacuum forming

### 13.2.1 Silicone Rubber Moulding

*Room Temperature Vulcanization (RTV)* silicone rubber enables fabrication of flexible moulds which can be used for manufacturing parts with intricate detail and undercuts. To produce a silicone rubber mould, the part model is first covered with a layer of sheet wax or modeling clay (approximately 6 mm thick) and placed in a box made in two halves. A backup material, usually polyurethane foam or epoxy resin is then poured into the box around the model and allowed to harden.

After curing of the resin, the two halves of the box are separated and the model as well as the wax or clay layer is removed to expose the two halves of the polyurethane mould. The model is then positioned in one half of the polyurethane mould and silicone rubber casting compound is poured in the space between the mould and the model. After curing (usually overnight) the other half is also cast and the two halves of the silicone rubber mould are put together for producing parts through injection moulding.

A different technique for producing silicone rubber moulds involves suspending the model in a box and pouring the resin all around the model. After curing, the mould is cut open along the desired parting lines to remove the model. The mould segments are then be mounted on an injection moulding machine to produce a few hundred plastic parts.

Silicone rubber moulds (Figure 13.2) have excellent chemical resistance, low shrinkage and high dimensional stability, making them suitable for producing parts in polyester, epoxy and polyurethane foam by injection moulding. They are however, more expensive than cast epoxy or laminated tooling, and therefore economical for small intricate parts or parts with undercuts which cannot be produced by rigid tooling. Such tooling are best when produced with wall thickness less than 12 mm.

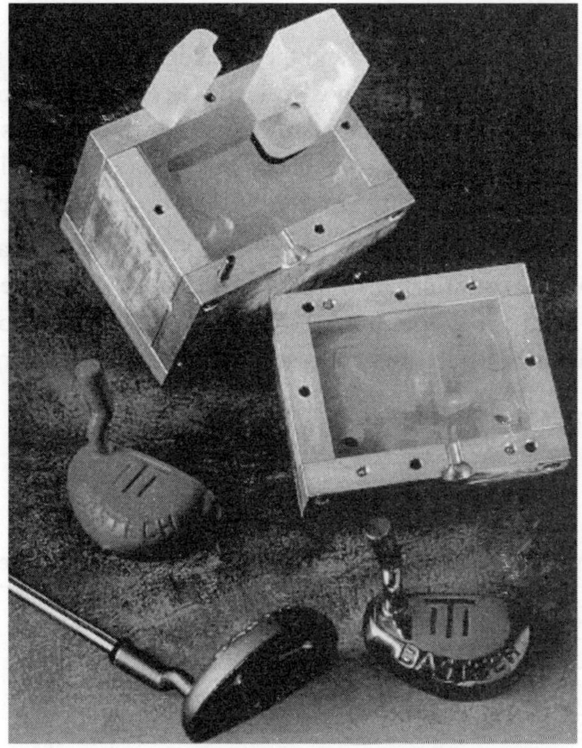

**Figure 13.2** Examples of indirect RT using silicon rubber moulding

Glass-filled acrylate resins are manufactured for plastic injection moulds. The mould can process about 180 ABS parts before heat distortion of the mould became unacceptable. Silicon rubber moulds, spin-casting, and vulcanization using a stereolithography master pattern are used to make multiple copies.

### 13.2.2 Epoxy Tooling

The process involves preparation of mould box and then strategically placing the master RP model. A release agent applied to the face of the mould box to facilitate easy release of the master model. An appropriate two component epoxy system is cast into the mould box. Fillers such as aluminum and steel are added to the resin to impart hardness, impact strength and wear resistance. After the curing time of epoxy the master model can be easily removed giving the necessary profile.

As shown in Figure 13.3, the Epoxy Tooling process starts with an RP master part, casts a silicon RTV (Room Temperature Vulcanizing) negative, casts a silicon RTV positive from the negative, and finally casts the aluminum filled epoxy tool from the RTV positive. The silicon RTV is used to aid in the separation process because its flexibility makes separation from the pattern less difficult.

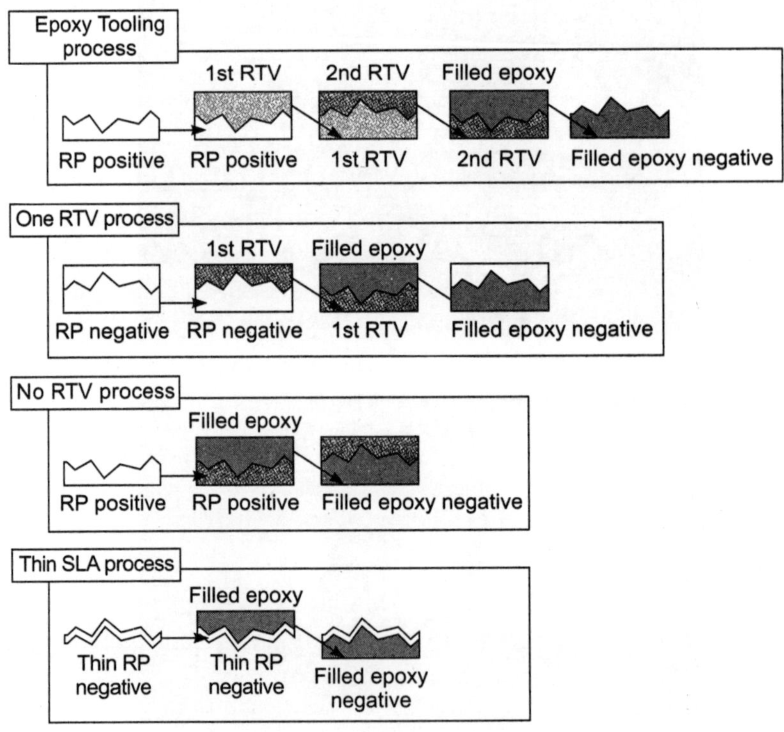


**Figure 13.3** Principle of epoxy tooling

Figure 13.3 shows three ways to reduce the number of steps required in the original process. The process that results in an aluminum filled epoxy tool with the fewest steps is the No RTV process. In the No RTV process, the filled epoxy tool is generated directly from the RP master part. Besides having the fewest steps, another benefit of the direct process is that it has the potential for high accuracy since accuracy is not lost in additional intermediate steps. Another alternative process that may be useful in some cases is to build an RP negative as a master. Using this alternative only a single silicon RTV positive is required. In general, however, the RP negative is more difficult to post-process (smoothing rough surfaces, cleaning, etc.) since the part features are inside a cavity. Still another possible alternative is to generate the external surface of the tooling in an SLA epoxy. Then, the shell is backfilled with the aluminum filled epoxy. The accuracy of this method is limited only by the accuracy of the stereolithography machine (SLA). Other benefits of this method include minimal SLA build time compared to generating the moulds in 100% SLA epoxy and minimal effort required after the use of stereolithography compared to the previous methods. However, a big disadvantage is that the mould is not likely to be as durable as a 100% aluminum filled epoxy mould.

Tooling is considered the most important application area for use of RP technologies. An injection moulding tool was fabricated directly from a 3 mm thick stereolithography shell that was later backfilled with aluminum shot and epoxy resin. This tool produced about 200 injection moulded automotive parts. The mould inserts showed signs of wear, chips on corners, and some pitting. Nevertheless, in cases where only a few parts are needed for design verification of a new component under development, this process is a viable way to make a tool directly from a computer model at a reasonable cost.

Two examples of epoxy tools are shown in Figure 13.4.

**Figure 13.4** Examples of indirect RT using epoxy tooling

### 13.2.3 Spray Metal Tooling

This process uses a high velocity electric arc metal spray generating system to deposit finely atomized particles of molten metal onto a model surface to create a metal shell mould (Figure 13.5). The metal is usually a low melting point metal or alloy of zinc, nickel, copper and aluminum.

Kirksite is one of the most widely used alloys for metal spraying. The metal wire is supplied from a spool and melted by an arc produced in a spray gun. The molten metal is propelled by a jet of compressed air, causing it to break up into fine particles which are deposited on the model surface. The model is coated with a release agent to anchor the initial spray coat, provide adequate temperature resistance and enable good detail besides facilitating separation of the metal shell from the model. The compressed air, arc power and metal feed rates are carefully optimized to produce a forceful spray which is cool (less than 100 °C) compared to the melting point of the metals involved. The metal shell thus produced, which is about 1 to 2 mm thick, can be reinforced by either layers or a solid mass of resin. The opposite sides of the part model can be sprayed to produce the two halves of the mould, which after reinforcement can be used for thermoforming, plastic injection moulding, blow moulding, sheet metal forming and vacuum casting (for making polyurethane or wax patterns).

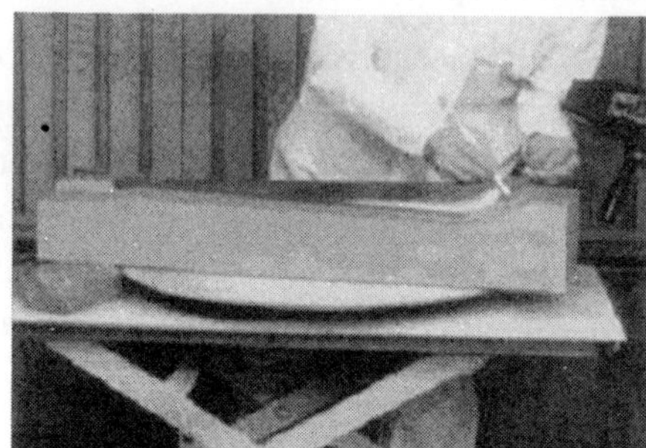

Preparing the model

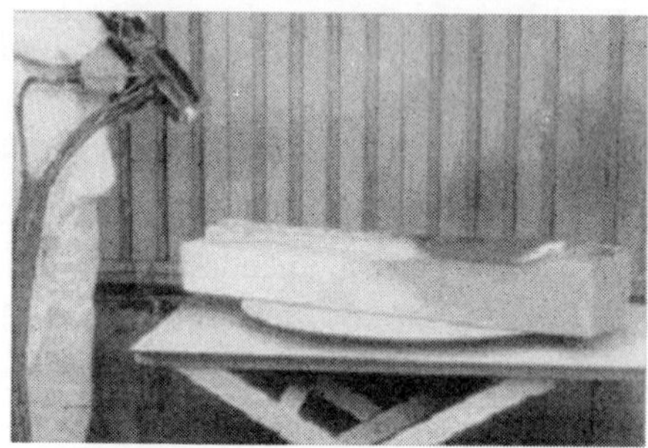

Cool spray the model with mouldmasking wire, and gun

Back-up the metal sprayed shell with high temp resin

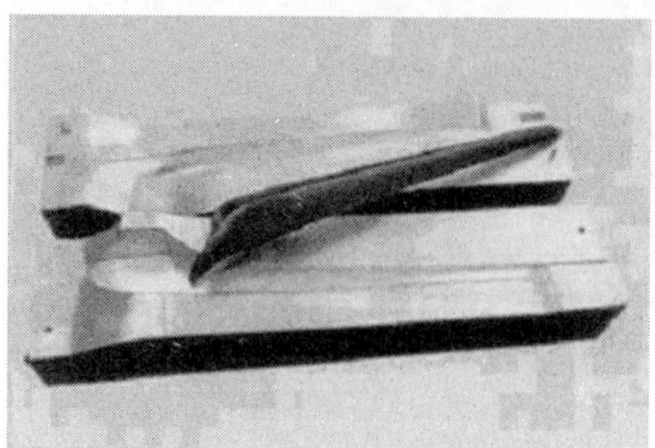

Finished mould in only 6 hours

**Figure 13.5** Principle of spray metal tooling

For metal spray method, size is no limitation, and this process has been successfully used for producing the tooling for sheet metal forming of automobile body parts. Plastic, wood or metal parts produced by conventional or RP techniques can be used. The porosity and low strength of sprayed metal moulds however, results in a shorter life compared to tool steel moulds. Specially developed copper and nickel plating techniques have been suggested for zinc sprayed tooling to improve the surface finish and prolong the tool life.

A similar method developed, involves simultaneous spraying of liquid metal droplets by metal arc process and bombardment of shot peening balls at the rapidly solidified metal surface. The shot peening not only improves the surface finish, but more important, relieves the stresses caused by rapid solidification of the molten metal. The resulting tooling has superior strength and toughness, and can produce larger number of parts during production. This method is called as *Manufacture Using Spray Tooling (MUST)*. Another recent development in metal arc spray tooling is a successful manufacture of steel tooling using plasma arc spraying on epoxy models. The tool obtained by this process is shown in Figure 13.6.

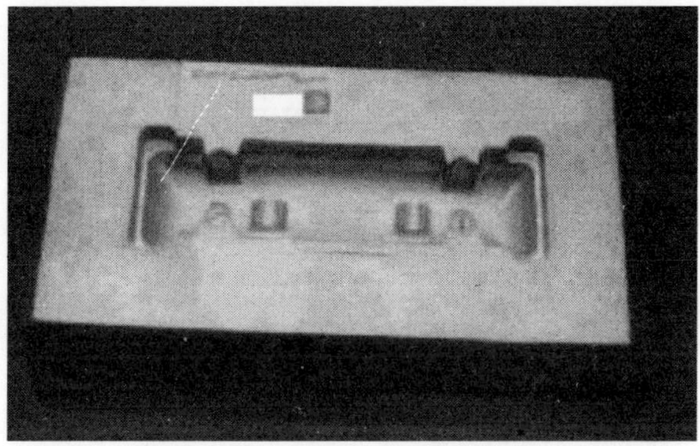

**Figure 13.6** Example of a tool made by spray metal tooling

Fused Deposition Modelling (FDM) process is used to make substrates. This process is akin to making the reverse of the desired part. Using an electroplating process that does not provide stress or heat, metal is applied on the backside of the part. After building a certain amount of metal on the back side of the part, it is back-filled with an epoxy material, aluminum shot material, or some other material to hold it so that it can be used as an injection mould, which has a 3 mm thickness of metal cladding interface between the mould and the back filling. A tool obtained using this method is shown in Figure 13.7.

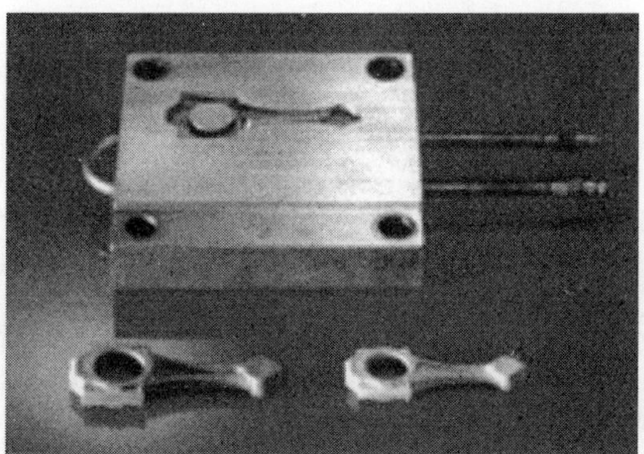

**Figure 13.7** Top: Tooling for injection moulding produced by FDM and electroplating the tool (mould half)
Left: green part (connecting rod); and right sintered part

### 13.2.4 Electroforming

This system uses plastic master patterns for the fast fabrication of nickel ceramic composite (NCC) tooling for medium volume plastic injection moulding runs. The method is based on plating nickel over the patterns and then reinforcing the thin, hard nickel face with a stiff ceramic material (Figure 13.8). Dimensional accuracies for the moulds are on a par with those of the original patterns (±0.1 mm).

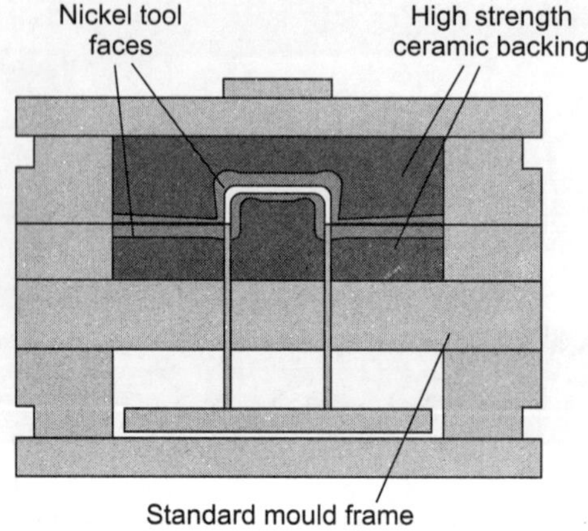


**Figure 13.8** Features of NCC tooling system

The model is coated with a conductive silver based material and placed in an electroforming bath of nickel sulfamate where a thin nickel layer is plated over it. The typical nickel plating thickness over the tool face varies from 1 to 5 mm. After electroforming, the model continues to serve as a fixture, holding the nickel shell in place. Alignment and accurate dimensions are maintained during the chemically bonded ceramic casting process by stabilizing the shell to the model. Prior to casting, cooling lines are custom fitted and fixed in place. The resulting precision nickel shell and model assembly is then attached to a standard pocketed steel mould frame using a high strength composite material. The ceramic is vacuum casting through a small opening in the back of the frame. After the curing for about a day, the opposing side is cast. The two halves are separated, the model is removed, and the mould is post-cured (Tool Figure 13.9).

**Figure 13.9** Tool made by Nickel ceramic composite using electroforming

### 13.2.5 Pattern for Sand Casting

Sand casting can be done from RP parts, by either the primary or the secondary route. Patterns can be made in hours using RP instead of days (as is the case with conventional processes) after giving the necessary allowances for shrinkage.

The primary route involves making sand moulds directly from an RP model such as the LOM model. The secondary route involves the following steps:

(a) Build polyurethane moulds from the RP model.

(b) Build polyurethane or epoxy pattern equipment (foundry tooling).

(c) Produce expendable sand moulds.

(d) Cast the metal parts.

Polyurethane and epoxy secondary tooling used for sand casting make a durable pattern material sufficient to produce hundreds of castings from a single set of tooling with a tolerance range of 0.5 – 1.0 mm. Typical functional parts manufactured using the RP sand-cast process include cylinder heads, manifolds (Figure 13.10), valve bodies, brackets and transmission housings.

**Figure 13.10** Automobile Manifolds made by Sand Casting

The LOM machine can be integrated into its manufacturing base, using it for fabricating complex patterns for sand casting. It is run on a 24 hour basis. Right next to it is a CNC (computer numerical control) wood milling machine. Depending on accuracy requirements, a company can choose between CNC machining and the LOM machine for pattern fabrication. CNC is used for greater precision, but the LOM machine is easier to use. Using the LOM machine has increased the production by 50%.

### 13.2.6 Pattern for Investment Casting

Investment casting, often called lost wax casting, is regarded as a precision casting process to fabricate near-net-shaped metal parts from almost any alloy. Although its history lies to a great extent in the production of art objects, the most common use of investment casting in more recent history has been the production of components requiring complex, often thin wall castings. The sequential steps of the investment casting process will be briefly described, with emphasis on casting from rapid prototyping patterns.

Many RP processes such as FDM, SLA etc. have wax as one of their materials. The patterns made using this method can be used for investment casting. An example is shown in Figure 13.11. Apart from wax, materials such as polystyrene (expended as well as unexpended) can also be used for these patterns.

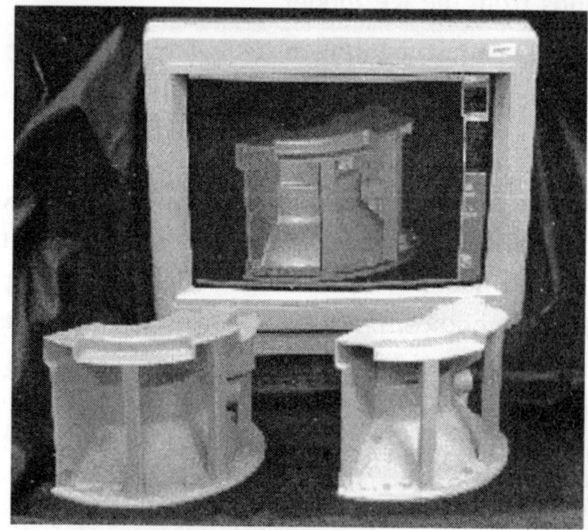

**Figure 13.11** CAD solid model (top), polycarbonate pattern using SLS process (left), A356 aluminum casting (right) from investment casting

### 13.2.7 QuickCast™ Process for Investment Casting

The ultimate aim in RP is to fabricate components in metal - the same material as the final parts. In particular, the focus is on investment casting, complex metal parts. The basic problem with using an SLA pattern for direct shell investment casting is the stress generated in the ceramic shell as a result of thermal expansion of the plastic pattern during heating. To tackle this problem, it is suggested that the patterns should be hollow so that they would implode instead of exploding. However, totally hollow patterns would bow in from hydrostatic pressure when dipped in the ceramic slurry. The process improvement that stemmed from this concept is the QuickCast build geometry, which produces an open lattice structure about two-thirds hollow. An internal triangular grid provides strength, while an outer skin provides pattern definition and smooth surface finish.

The QuickCast pattern, meeting the above requirements, is made from Epoxy Resin on the SLA machine. The pattern is then gated and prepared for dipping in a container of fine "face coat" ceramic slurry to form the casting shell. After drying, several more ceramic shells are added, and the assembly is placed in an aspirated oven and fired at higher than 980°C for an hour, which cures the shell and burns out the epoxy resin pattern. The ceramic shell is now ready for pouring the molten metal. Figure 13.12 shows a knee implant made using QuickCast.

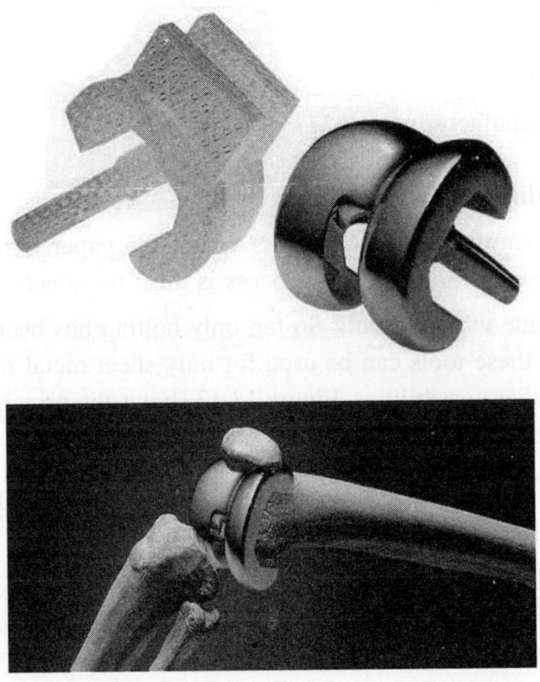

**Figure 13.12** Knee Implant made by QuickCast process

Aviation companies use RP patterns for investment casting. Designing and manufacturing complex metal castings using 3D System's QuickCast$^{TM}$ build style to fabricate patterns is possible. Foundry process parameters have been developed for successfully casting RP patterns. RP castings can be used in testing prototype flight hardware.

A new four-cylinder engine block using CATIA solid modelling software has been developed. The solid model design of the full size engine block is sent to 3D Systems for pattern fabrication. The completed investment casting pattern is used for investment casting. The completed aluminum castings result in a significant savings in time and cost.

3D Systems use stereolithography machines to build QuickCast epoxy patterns for investment casting.

## 13.3 DIRECT RAPID TOOLING PROCESSES

Methods of rapid prototyping that can be used directly to fabricate metal objects include:

- Laminated Tooling
- Selective Laser Sintering

- Shape Deposition Manufacture (SDM)
- Laser Deposition
- Droplet Deposition
- Hybrid Layered Manufacturing (HLM).

### 13.3.1 Laminated Tooling

Laminated tooling is very similar to LOM process where the paper is replaced by metallic sheets. There is no automatic machine available. This process is ideal for sheet metal tools.

Joining the layers is some what difficult. So far, only bolting has been found satisfactory. Due to local joints in this method, these tools can be used for only sheet metal forming dies. Bonding using adhesives and brazing are other possibilities. Figure 13.13 shows the principle of laminated tooling and Figure 13.14 shows a laminated stamping die and a part made using that.

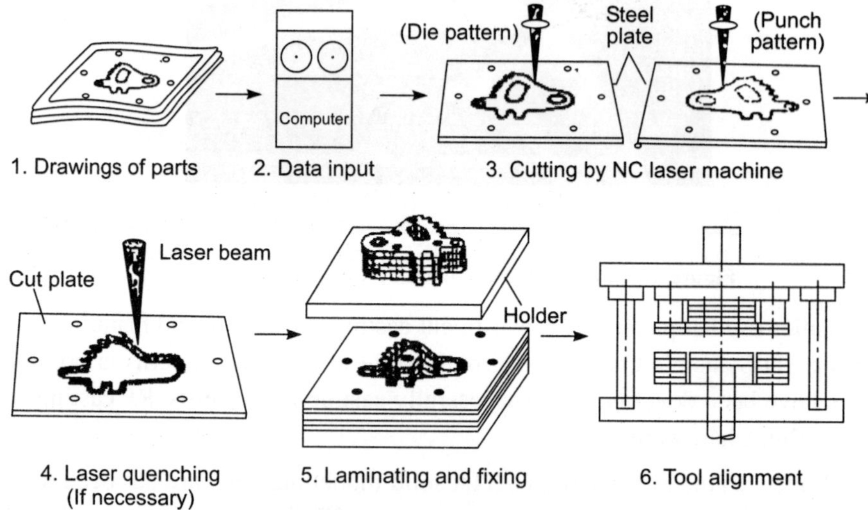


**Figure 13.13** Principle of laminated tooling

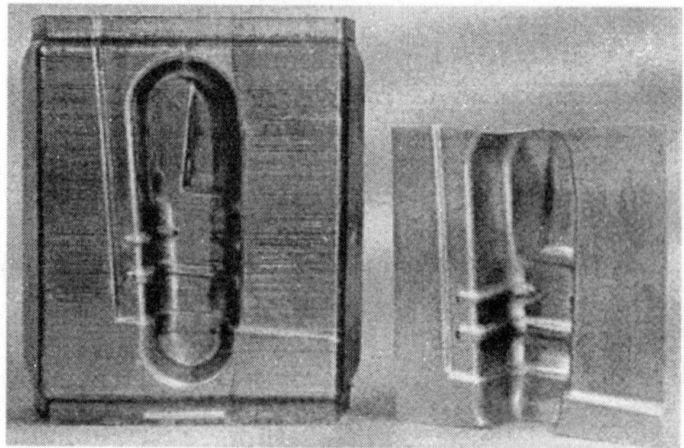

**Figure 13.14** Laminated tool of a stamping die (left) and the part produced using that (right)

### 13.3.2 Selective Laser Sintering

Selective laser sintering (SLS) is a rapid prototyping process first developed at the University of Texas, U.S.A. In this process, metal powders are coated with a thermoplastic binder. These coated powders are then selectively fused together with a laser in the SLS process. This bonds the metal powders together to form mould components represented by a CAD file, thus producing a green part. The green part is then post-processed in a furnace, where the binder is burned out, and the metal powders are bonded together through traditional sintering mechanics. This part is now referred to as a "brown" part; it exhibits geometry but is also porous in nature. The brown part is then infiltrated with a second metal to form a fully dense mould. Figure 13.15 shows a mould core and cavity set that was created with this process. The properties of this mould are similar to those of 7075 aluminum.

**Figure 13.15** Die inserts made using SLS process

Some SLS equipment uses a high power laser and material of low melting point and hence they do not use binder coating.

Apart from permanent metallic moulds, SLS (and 3DP) are capable of building sand moulds without any patterns.

### 13.3.3 Shape Deposition Manufacturing (SDM)

SDM first deposits layers as near-net shapes and then machines them to accurate dimensions before additional material is added. The basic steps for building parts with Shape Deposition is depicted in Figure 13.16. To form each layer, the growing shape is transferred to several processing stations. First, the material for each layer is deposited as a near-net shape using a novel weld based deposition process called microcasting (Figure 13.16a). The part is then transferred to a shaping station, such as a 5-axes CNC milling machine, where material is removed to form the net shape (Figure 13.16b). In the next step, the part is transferred to a stress relief station, such as shot peening, to control the build up of residual stresses (Figure 13.16c). The part is then transferred back to the deposition station, where complementary shaped, sacrificial support material is also deposited. Each of these operations is described in detail below:

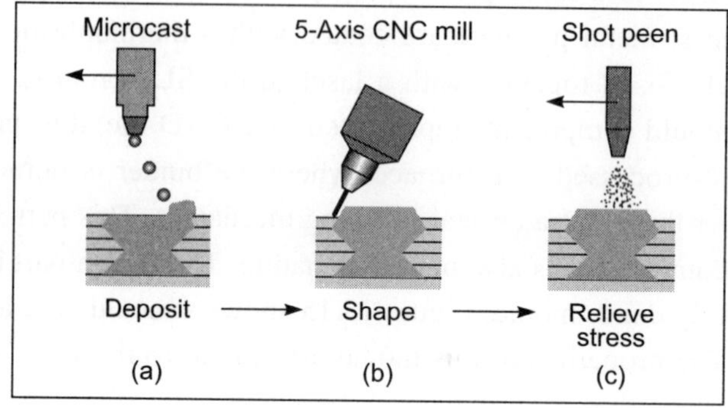


**Figure 13.16** Principle of SDM: Creation of a layer using SDM

Shape Deposition decomposes the CAD model of the part into slices which maintain the full three-dimensional geometry of the outer surface. The total layer thickness and the sequence for depositing the primary and support materials depends upon the local surface geometry. Consider the shape in Figure 13.17 which represents three fundamental features which can be found in a layer; non-undercut (relative to the build direction), undercut, and a combination of both.

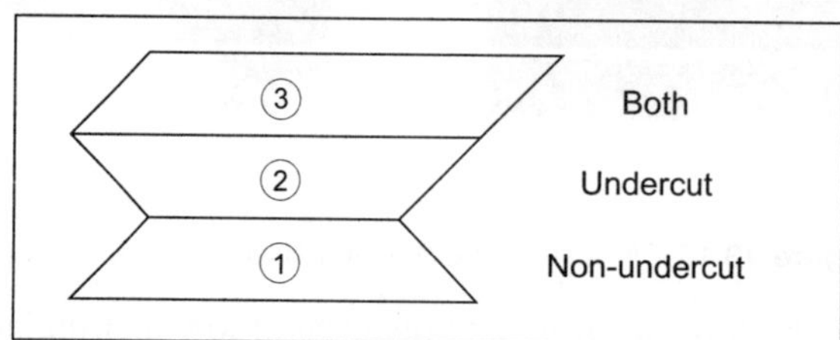


**Figure 13.17** Principle of SDM: Cross-section of example shape

This shape can be formed as follows:

- In the first layer, which contains only non-undercut features (Figure 13.18), the primary material is deposited (Figure 13.18a) and shaped (Figure 13.18b) first. This layer is completed by depositing the support material (Figure 13.18c) and planing the top surface (Figure 13.18d).

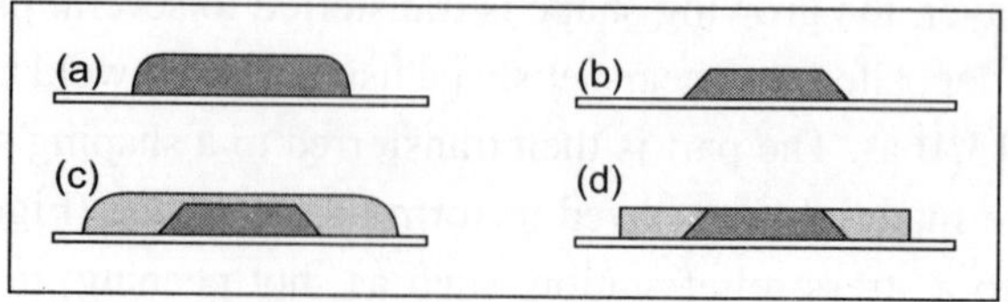


**Figure 13.18** Principle of SDM: Manufacture of non-undercut features (First layer)

- The second layer, which contains only undercut features (Figure 13.19), is created by depositing (Figure 13.19a) and shaping (Figure 13.19b) the support material first. This forms a moulding cavity into which the primary material is then deposited (Figure 13.19c) and the layer is finished by planing the top surface (Figure 13.19d).

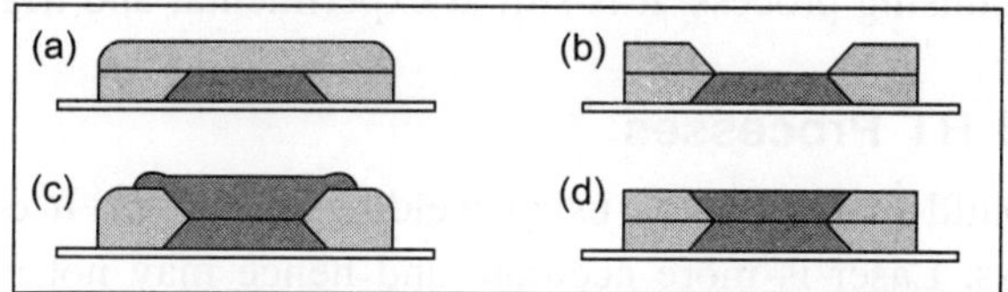


**Figure 13.19** Principle of SDM: Manufacture of non-undercut features (Second layer)

- For the third layer, the support material must be subdivided (Figure 13.20). The section of the support material with no undercuts is deposited and shaped first (Figure 13.20a). Next, the primary material is deposited and the non-undercut surfaces are shaped (Figure 13.20b). Finally the remaining portion of the support material is deposited and the layer is planed (Figure 13.20c). In general, for layers containing a combination of undercut and non-undercut surfaces, the individual materials have to be split into smaller segments. Each segment contains undercut surfaces only in those areas which are adjacent to previously deposited segments of the layer.

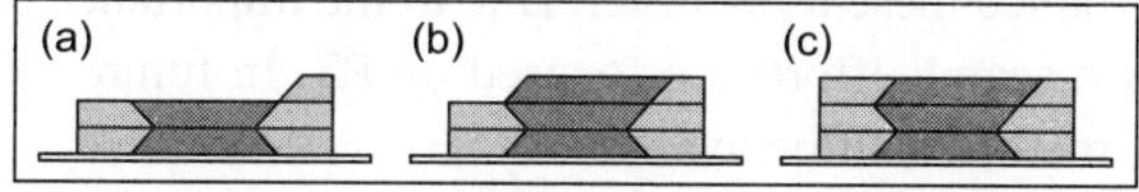


**Figure 13.20** Principle of SDM: Manufacture of arbitrary layers (Third layer)

Figure 13.21 shows a comparison of cross-sections of the part manufactured with other RP techniques and SDM. SDM requires very less number of layers as it uses adaptive slicing of 1$^{st}$ order edge approximation.

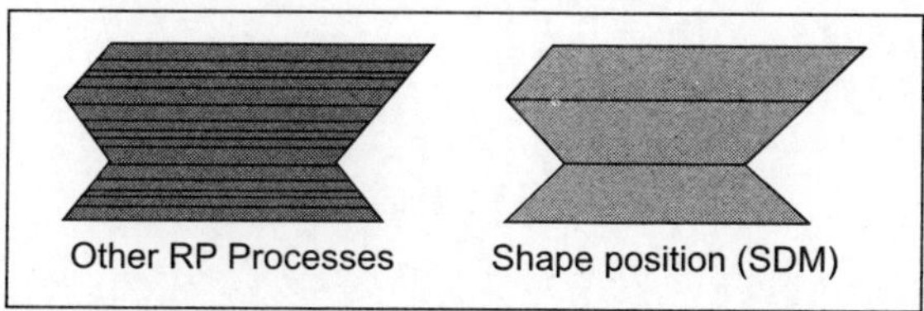


**Figure 13.21** Comparison of the layers of SDM with the layers of other RP processes

SDM system has also been utilized to fabricate an advanced, multimaterial injection moulding tools. The tool has copper deposits and cooling channels to minimize cycle time and Invar core to reduce distortion due to residual thermal stresses. The outside of the tool is stainless steel to prevent corrosion while in the acid bath used to remove the copper used as a sacrificial support material. The Invar and stainless steel were deposited in a continuous, outward spiral, where the transition to stainless steel was made within the outer few passes along the contour. The same technique could be also used to produce long life, wear resistant tools with a hard, corrosion resistant outer shell. This technique can be extended to any multimaterial application where stress concentrations due to sharp material interfaces may limit performance. Multiple materials also offer the promise of building integrated components in one operation.

Although SDM is a promising process, it is still at experimental and trial stage.

### 13.3.4 Welding-Based RT Processes

Several efforts are on to build metallic parts using welding processes. It could be laser welding or MIG/MAG welding process. Laser is more accurate and hence may not require machining. When MIG/MAG is used, it is invariably a hybrid process where the welding is used for near-net layer deposition and the required accuracy is obtained by machining.

When welding is used, no support material can be used that does not stick. Although copper has been tried as a support in SDM, the surface quality at the interface is not good. Therefore, presently these hybrid processes are restricted to tools only. Tools are inherently free from overhanging features and hence do not require any support.

## 13.4 CONCLUSION

The manufacture of the production tools takes several months and they also take up a major part of the investment. Therefore, it is important to produce these tools faster even if they will have lower life. Rapid Tooling will help to produce these tools faster. Due to the importance of cost and time reduction in tool manufacture, a lot of research efforts are focused on RT. In future, direct RT processes which use the sophisticated slicing method will be available.

❑❑❑

# Part IV

## SYNERGIC INTEGRATION

# CHAPTER 14

# Concurrent Engineering

## 14.1 INTRODUCTION

*Concurrent Engineering (CE)*, also known as *Simultaneous Engineering (SE)*, as these names suggest, is the approach of doing all the activities at the same time as far as possible. It is the unison of all the facets of product development and life cycles to minimize modifications in the prototype, i.e., to decrease the design iterations performed during the product design.

CE has been previously used in electronic and manufacturing system design. The objective of CE is to reduce the system/product development cycle time through a better integration of activities and processes. A more formal definition has been provided by the Institute for Defence Analysis (IDA), U.S.A.

"*Concurrent Engineering is the systematic approach to the integrated, concurrent design of products and related processes, including manufacturing and support. This approach is intended to enable the product developers to consider all elements of the product life cycle from conception to disposal, including quality, cost, schedule, and user requirements*".

The basic tenet of CE is the integration of methodologies, processes, human beings, tools and methods to support product development. CE is multi-disciplinary in that it includes aspects from object-oriented programming, constraint programming, visual programming, knowledge-based systems, hypermedia, database management systems and CAD/CAM.

CE involves the interaction of diverse groups of individuals who may be scattered over a wide geographic range. To enable effective and complete communication among them, there are certain technological concepts that must also become organized into concurrent layers. Distributed information sharing and collaborative/ cooperative work are important techniques to maintain and exceed the current level of software development productivity. CE takes advantage of shared information and allows simultaneous focus on different phases of the software development life cycle. Many existing World Wide Web (WWW) capabilities could support a wide area CE environment.

CE is a strong, dependable product development methodology that can be continuously upgraded and modified. It is a methodology that leads to a significant reduction in cost and development time without sacrificing any of the desired product specifications. Moreover, it is simple to comprehend, easy to implement and easily adaptable to a diverse nature of product development activities. Now CE gets its greatest exposure in the industries. Companies have changed their work procedures by creating teams of people and giving them, rather than departmental managers, the power to make changes to improve timeliness, to increase employee competence, to enhance quality, to reduce waste or to increase yield. This generally is accompanied by the heightening of consciousness for quality in one's work and by a boost in morale. The end result shows up in the profits because customers are pleased by the change in the company's responsiveness, and by the discernible improvement in the quality of its product or service. Some of the companies are large engineering firms while others are small ones operating in a limited market. There are firms in the service industry who have demonstrated the same ability to improve themselves and to overcome the old hierarchical flows of work and information in their companies. By pushing down decision-making to the lowest level at which the problem is encountered, and by making sure that the necessary institutional knowledge, no matter where it lies in the organization, is exploited to solve the problem, a new energy has been released.

## 14.2 BENEFITS

CE-based design is being introduced in a number of manufacturing organizations. Early experiences have produced several emerging benefits. The cost of design is actually a fraction of the total cost of product life cycle. Design, however, creates the obligation of expenditures over the product life cycle, for the spending that will be incurred. In other words, a major part of the product cost is fixed at the design stage.

There are many positive effects of CE design. CE-based design has emerged as a response to several interrelated events described below:

- *Limitations of MRP*: The lack of success of classical "MRP" extensibility for complex product manufacturing beyond its original material management & scheduling scope.
- *Generation of new ideas Creating Conflicts*: New ideas and concepts to improve manufacturing into a consistent framework or architecture run into difficulties.
- *Technology Push*: The emergence of the powerful technical workstations, networked communications and highly capable software & systems prevalent in the computer industry.
- *Business Infrastructure Evolution*: The rapid disappearance or demise of industrial corporations and the emergence of "global competition" and the reasons behind the rapid business changes. These changes include the need to:
  - Constantly reduce product costs
  - Substantially shorten time to market and competitive response times
  - Constantly improve product quality.
- *Business Management under Competitive Pressures*: The recognition that the successful 21$^{st}$ century organization will look substantially different from that of the post.

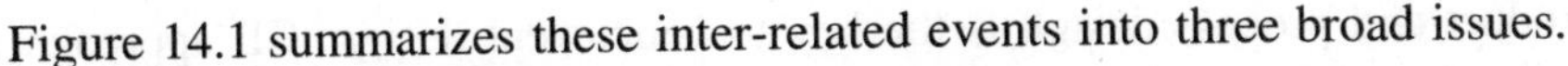

Figure 14.1 summarizes these inter-related events into three broad issues.

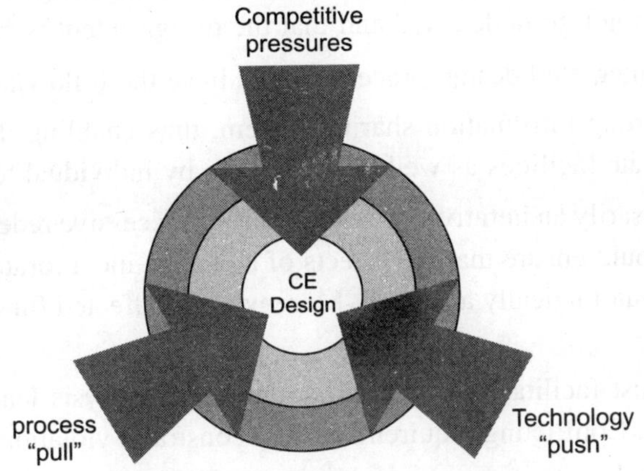


**Figure 14.1** CE-based design pressures

## 14.3 METHODOLOGY OF CONCURRENT ENGINEERING

CE is characterized by a focus on customer's requirement. Moreover, it embodies the belief that quality is inbuilt in the product, and that it (quality) is a result of continuous improvement of a process. This concept is not new; in fact, the approach is quite similar to the "product team" approach characteristic of small organizations. The "product team" essentially comprises of a small group of people working closely for a common endeavor, which might be product development. The magnitude of the problem is usually small with few conflicting constraints. The approach works well for small organizations; however, in large organizations the technique needs to be modified and restructured. It is here that CE comes into picture. CE envisages translating the "product team" concept to big organizations which will work with a unified and focussed product concept. As the team members can be at geographically different, networked locations, this requires far-reaching changes in the work culture and ethical values of the organization.

The methodology of CE, which is discussed here, can be possible only if the traditional work culture (existing in most present day enterprises) paves the way for a more open, equitable and trustful environment, where the hierarchical position of an individual is not a barrier in information exchange. The design group can effectively utilize the CE approach only when it works cooperatively and consistently. The group should have the capability to analyze individual viewpoints and reach a common decision and technological agreement.

The objectives CE are:

- Elimination of waste
- Reduction in lead-time for product development and product delivery
- Improvement of quality
- Reduction of cost
- Continuous improvement.

Integrated, parallel product and process design is the key to concurrent design. CE approach as opposed to the conventional sequential approach advocates a parallel design effort. The objective is to ensure that serious errors do not go undetected and that the design intent is fully captured.

The above mentioned integrated design process should have the following features:

- There must be a strong information sharing system, thus enabling the design teams to have access to all corporate facilities as well as work done by individual teams.
- Any design is necessarily an iterative process requiring successive redesigns and modifications. The CE process should ensure that the effects of a change incorporated by one team on other design aspects are automatically analyzed. Moreover, the affected functions should be notified of the changes.
- The CE process must facilitate an appropriate trade-off analysis leading to product-process design optimization. Conflicting requirements and constraint violations must be identified and concurrently resolved.
- All the relevant aspects of the design process must be recorded and documented for future reference.

In order to achieve the above, a CE environment will have the following three components:

(i) Data architecture

(ii) Information framework

(iii) Software tools and services.

These are discussed in the following sections.

### 14.3.1 Data Architecture

The primary purpose of data architecture is to organize the design and manufacturing knowledge of the enterprise to accomplish maximum exploitation of this database. Due to the very nature of concurrent engineering, the diversity and quantity of data is immense. Therefore, only its structuring can result in rapid access, global transparency and fast information transfer.

Some main features of the data architecture are enumerated below:

- Creation of data base packages that will enable an exhaustive storage of information on product-process descriptions and corporate resources.
- Providing sufficient flexibility to allow easy modifications of the database. Moreover, there should be the capability of recording the "by whom" and "why" of the modifications done.
- The information must be organized into various sub-classes with explicit relations amongst them. These classes can be thought of as pertaining to: life cycle phase (design, manufacturing, support etc.), requirement attributes (performance, cost, time factor etc.), product type etc. Provision for extension of existing classes and addition of newer ones must be made.
- There must be a mechanism to ensure that the contents of the database are quickly and conveniently accessible to the various product development teams.

- The databases must be strongly protected against unauthorized use, and only identified users should have access to the system.
- There must be an extensive index to the contents of the database, record of applications software, available computing facilities and authorized users.
- There should be a facility for detailed documentation of an ongoing design effort. The idea should be not only to record the design data but the method of design as well. The method used for designing a present product can prove to be very valuable to future designers tackling a similar problem.

Figure 14.2 illustrates the architectural planning process discussed above.

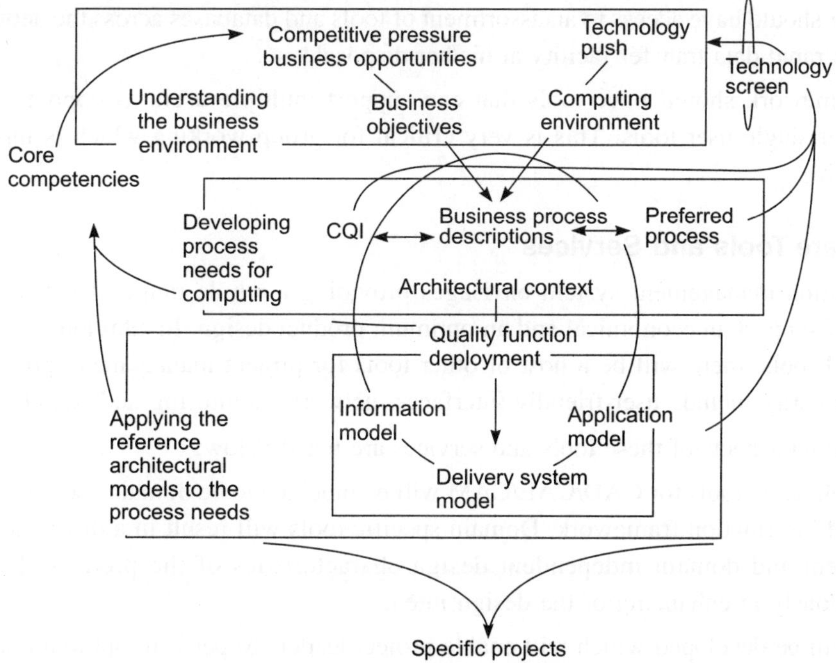


**Figure 14.2** Architectural planning process

### 14.3.2 Information Framework

CE is dependent to a great extent on information network for managing the communications between various groups involved in product life cycle activity design. There is a need for an information framework capable of accomplishing the cooperation and coordination necessary for concurrent engineering.

Some of the key features of the framework are:

- The framework should be based on a set of standards and specifications so that it is general purpose. At the same time, however, it should have attributes of adaptability to the special needs of an enterprise. Diligent efforts are required for developing a standard framework for

CAD/CAM/CAE application tools. Moreover, the framework should be compatible with the existing hardware and software platforms.

- The framework should serve as a common environment for designers to perform a cooperative design effort. It should facilitate parallel execution of design activities.
- There should be mechanisms to highlight conflicts and notify designers of decisions and modifications. In addition, the framework has tools to analyze and evaluate global effects of local design modifications.
- The framework should have intelligent information filtration and interpretation mechanisms so that experts of one domain can derive maximum benefits out of knowledge pertaining to other domains.
- The user should have access to an assortment of tools and databases across the network. He/she also has rapid data transfer facility at his/her disposal.
- The framework should have tools that can support multiple users as opposed to currently prevalent single user tools. This is very critical for group working which is inherent to CE approach.

### 14.3.3 Software Tools and Services

The CE information management system envisages providing a whole range of software tools and services that will support an economical and an optimum product design. In addition to a multitude of CAD/CAE/CAM tools, there will be a host of other tools for project management, process planning etc. Services primarily include user-friendly interfaces, network monitoring and Rapid Prototyping.

Some important aspects of these tools and services are listed below:

- The application tools for CAD/CAE/CAM will be modeled to be domain specific and based on standard information framework. Domain specific tools will result in a distinction of domain dependent and domain independent design characteristics of the product. This will help tremendously in enhancing of the design intent.
- Tools will be developed which will enable project leaders to perform optimum task planning, monitoring and scheduling for timely project completion.
- User-friendly interface modules will be implemented and this will enable the user to work in a fully interactive mode. Moreover, the users will have access to all tools from their individual workstations.
- A facility will be provided to keep account of the interactions amongst product developers on the network. This will monitor the information flow on the communication channels and diagnose faulty sub-systems.
- A "first unit production" method will be established which will produce a prototype of the designed product with the desired quality requirements. Feature based design representations will be prepared which will be translated to tooling and manufacturing information. This information would be utilized by CNC (computerized numerically controlled) machines and robotic systems to perform actual manufacturing of the prototype.

## 14.4 INTEGRATION IN CONCURRENT ENGINEERING

CE is a systematic approach to the integrated, concurrent design of products and their related processes, including manufacture and support. This approach is intended to enable the developers, from the outset, to consider all elements of the product life cycle from conception through disposal, including quality, cost, schedule, and user requirements. Thus, systematic integration of methodologies, processes, human beings, tools, and methods is one of the basic directions of CE. This integration and developing tools that better accommodate, on the one hand, the iterative and evolutionary and, on the other hand, the concurrent nature of product design forms the foundations of CE systems. CE systems represent the evolution of a number of distinct paths of technological development including object-oriented programming, constraint programming, visual programming, knowledge-based systems, hypermedia, databases and information retrieval, and CAD/CAM/CAE. These technologies have reached a stage of maturity on their own so that it is now possible to define an overall unifying structure for product design. This structure caters to the needs of product developers for information that is relevant, easy to obtain, and helpful. The architecture of CE systems is considered as four levels consisting of an object-oriented database, an intelligent data base engine, a high-level interface and high-level tools. Within this architecture the integration of humans, their tools and methods, and processors to form a CE environment for the development of products is treated in the context of cooperating knowledge bases.

The CE system architecture forms a new discipline, which is responsible for providing the functionality of ready access to information throughout designing and manufacturing organizations. The CE data base four-level architecture is illustrated in Figure 14.3.

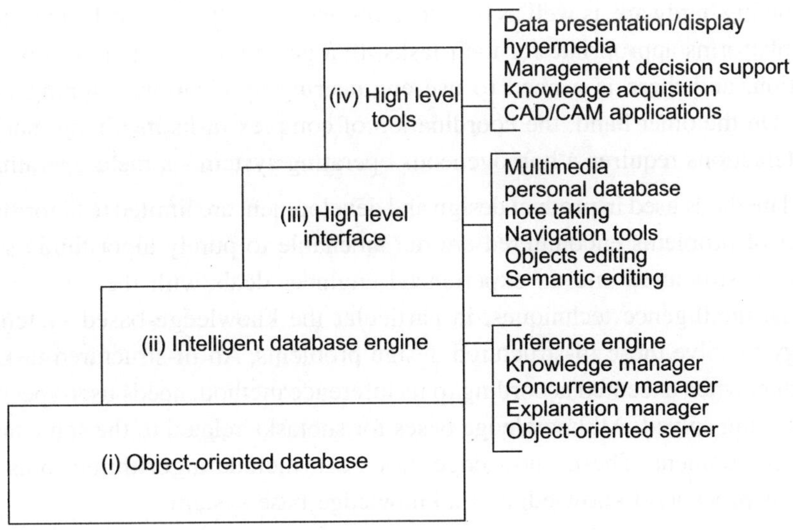


**Figure 14.3** CE system architecture

(i) The purpose of the *object-oriented database* in CE, considered as the bottom level of the four level architecture, is to provide an engineering environment that supports real-time inspection and modification of enterprise data.

(ii) The second level, above the object-oriented data base level, is the *intelligent database engine.* The engine consists of an inference engine, knowledge manager, concurrency manager, explanation manager, and object-oriented server.

(iii) The third level is the *high-level user interface*. This level creates the model of the task and data base environment that users interact with. Associated with this level is a set of representation tools that enhance the functionality of the engineering environment. The user interface is presented in two aspects. There is a core model that is presented to the user, and this model is based on hypermedia and visual programming techniques. In addition, there is a set of high-level tools which, although not an essential part of the core model, nevertheless enhances the functionality of the CE system for certain classes of user.

(iv) The fourth of these levels is the *high-level tools*. These tools provide the user with a number of facilities such as intelligent search, data presentation, data quality and integrity control, and automated discovery. They usually appear and work much as their stand-alone equivalents, such as spreadsheets and graphic representation tools, but they are modified so as to be compatible with the object-oriented database and knowledge-base models. The CE system high-level tools contain CAD/CAE/CAM applications and tools of knowledge representation environment. On this level, we also consider intelligent system design tools that provide facilities for designing the object-oriented database. These tools are somewhat peripheral to the central idea and functioning of the CE intelligent database, but they are needed as part of the product development process.

The above four levels pertain to software integration. In addition to that, a CE environment requires integration of computing hardware as well as people, tools and methods. Networked product developers may use different platforms appropriate for their tasks. In a general case, one developer can use more than one workstation, and there is a need to use engineering applications running under different operating systems. On the other hand, the coordination of complex tasks involving many humans and a long series of interactions requires a homogeneous operating system - a meta-operating system.

Many tools and methods used in product design and development are limited to algorithmic solutions. However, a number of problems encountered are not amenable to purely algorithmic solutions. Such problems are often ill structured and an experienced engineer deals with them using judgment and experience. Artificial intelligence techniques, in particular the knowledge-based systems technology, offer a methodology to solve these ill-structured design problems. An ill-structured task is defined by a knowledge base that, when executed according to its inference method, needs user-specific knowledge relevant to the task being solved. All knowledge bases for subtasks related to the same task form a task knowledge-based environment. These knowledge bases can include algorithmic tools and methods, which can be used as procedural knowledge of a knowledge-base system.

## 14.5 CE TRANSACTIONS

The generic services provide the team-working glue for the CAD/CAE/CAM and *Design for X (DFX)* Tools. The architecture of a CE environment, from this point of view, consists of the set of CE services and their interfaces. In the aggregate they serve to elevate solo designers employing single user tools in a personal workspace, to the level of team designers using remotely communicating tools on a network and sharing information in a common database. The introduction of group working as the norm produces so many requirements that the tools themselves will have to change their interface to the user and the network and the database in future. They will need to become much more open, allowing everything from extensions of user interfaces to remote tool access by simple end-user programming. These interactions are absolutely necessary for a reasonably comfortable level of integration of the tools into the CE environment. Besides, they must uniformly embrace the client-server model of distributed computing and conform to standards for co-existence such as X-Windows.

There are many generic services associated with the transactions the team members can effect with their assistance. Here is a list of CE transactions made possible:

- Query on the schema of an object
- Query on the existence and instances of an object
- Make a new instance of an object whose schema is defined
- Create a new version of an object
- Add a document to the design notebook, link it to specified nodes, and index it by certain keywords
- Get a document from the design notebook via its indexes
- Get a document from the design notebook navigating via the links from a previous document
- Get a document from the design notebook navigating via the links from the product data model
- Send a bundle of assertions to the project coordination workspace
- Send a task to the project coordination team dissemination
- Get specified attribute-value pairs from the product data structure of the coordination workspace, containing proposed but not yet agreed decisions
- Publish a file, i.e., place the data somewhere for wider access, inform the project coordination team of its location, and let it alert those users dependent on that data
- Send an "Alert" message to specific team members
- Define and configure a new assessment method for the product data
- Evaluate a previously configured assessment
- Produce a periodic report of multiple assessments, based on an event trigger or a data sufficiency condition
- Transmit data from one application to another or to a human user or to a generic service, such as the project coordination team
- Send an on-screen message directly to a logged-on user
- Meet synchronously with the team employing multiple media for exchange over the network
- Archive team exchanges
- Send a message of predefined type to an application

- Invoke an application anywhere in the network with given arguments, possibly transferring a file as argument
- Enquire about a resource
- Execute a set of programmes on several computers in a defined network sequence connecting the output files of one programme to the inputs of the others as specified
- Translate a file from one format to the neutral format and vice versa
- Capture on disk a part of a screen and/or transmit it to a logged-on user's screen in a window
- Add or modify a requirement or constraint
- Evaluate a requirement, executing a local or remote programme for the purpose, and returning the result of the evaluation
- Retrieve status of a set of requirements, by ownership, criticality or other characteristic.

These transactions fall broadly into 6 classes:

(i) *Look-up*: Information is scattered throughout an organization and stored in various media ranging from file cabinets to knowledge bases. What is needed here is an *Information Server* which functions as the *single point of inquiry* for any information need. Current developments in the area of object-oriented databases can form a solid foundation for the realization of this goal.

(ii) *Computation*: The information, which is gained by look-up or other means, is subjected to computation to add value. The main obstacle for *computation at will* is the heterogeneous and distributed nature of computing equipment and software applications. *Distributed yellow pages* of the hardware and software resources of an organization will potentially place the entire computational repertoire of the organization at an individual user's disposal.

(iii) *Communication*: Sharing information is the key to effective CE applications. However, the low band-width of the present networks, and the need to treat each media (text, graphics, voice and video) separately, have proved to be limiting factors in realizing effective and spontaneous communication. Present initiatives such as the gigabit networks, multimedia integration, and *meeting-on-the-network***,** will soon make it possible for each member of a *virtual team* to be *in-conference* with any other member at will.

(iv) *Negotiation*: CE is predicated on the ability of each virtual team member to negotiate with the group and reach consensus. The essential prerequisites for this are the global visibility of the evolving design, and the recognition of constraint violations brought about by individual decisions. Advances in technologies such as meeting-on-the-network, constraint management and project co-ordination via blackboards, point the way for reaching consensus on the network rather than solely through face-to-face meetings.

(v) *Decision Making*: In the course of product development, various decision-making tools will be enlisted by different members of the team. However, most of the present decision support tools are *single-perspective centered*. Advances are needed in the area of group decision support. Design assessment tools and *Quality Function Deployment (QFD)*, with the added capability for handling imprecise qualitative measures, are some of the developments that can assist in group decision-making.

(vi) *Archiving (Storing for retrieval)*: The *intellectual enterprise* that realized a product ought to be captured and exploited subsequently during re-design or development of related products. However, capturing design rationale seems to be one of the weakest aspects of present-day engineering activities. A hyper-media based design capture system coupled to the virtual team support environment will be needed to truly capture the multiplicity of product development decisions. Work on electronic design notebooks is directed toward fulfilling this important need.

Indeed, these six operations may be viewed as the basic *instruction set* for a CE environment.

The Product development process goes in stages. Each stage comprises steps shown schematically in Figure 14.4. The steps shown in the shaded blocks are conducted using the CE services and the transactions they make possible on the network without the product developer leaving his/ her workplace. The transactions mentioned above could be matched against the actions of the shaded blocks.

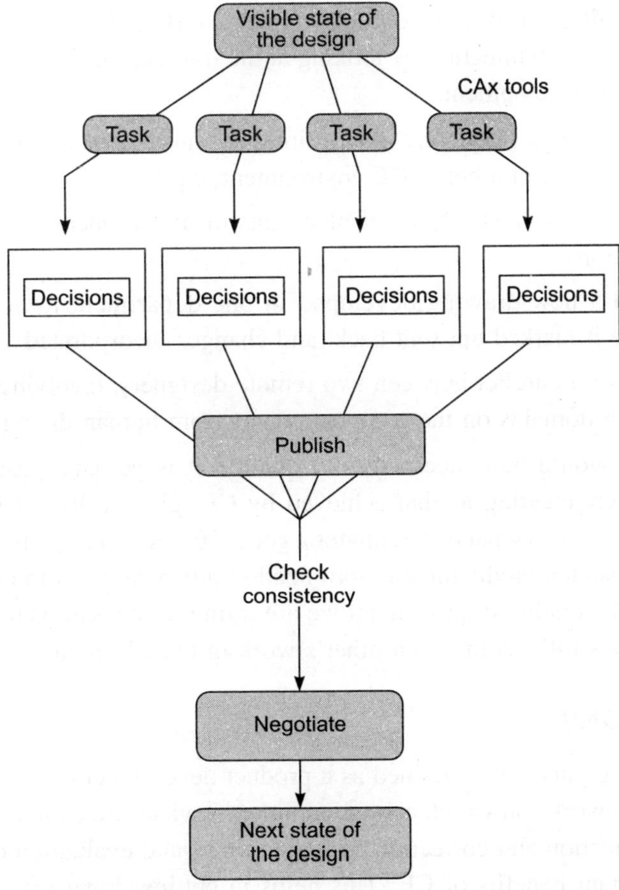


**Figure 14.4** Product development process in CE

The CE environment exists precisely to make possible the transactions described above between remote engineers, their engineering tools and the CE services themselves. These are transactions that would otherwise be performed, with or without computers and networks and databases, but with a substantial burden of bookkeeping, hand massaging of data, and conveyance of product and process knowledge via human minds.

### 14.5.1 Metrics of the CE Environment

The ability to carry out these representative transactions signifies progress toward a truly integrated CE environment. But it is not only the fact that such transactions can be performed, but the ease with which an organization can develop the inter-operation between engineering tools and CE services, and the simplicity with which engineering user can employ them, that is a measure of the integration. Integration is not an end in itself to be met at all costs. If the cost in terms of coding effort and information accessibility is very high, it is better not to do that integration by computer, but to break it down into a set of more primitive transactions and perform them with human intervention.

The effectiveness of the CE environment must be measured ultimately in terms of cost reduction, time reduction, and quality improvement of products created. But one can arrive at intermediate measures of the effectiveness of this architecture by looking at the transactions and gauging their contribution to the progress of product development.

One could choose certain frequently done primitive transactions and measure the elapsed *time* in the present environment and in a better CE environment, e.g.:

- Time to access and make copies of blueprints of all instances of a particular part satisfying certain conditions
- Time to send a new drawing developed in one department to another person in a remote location, have it marked up, sent back, and changes incorporated
- Time to resolve a conflict between two remote designers, involving negotiation and trade-off studies in both domains on the computer, apart from human dialogue.

Another measure would be concerned with quality. It is perhaps possible to achieve the same quality by sequential engineering as that achieved by CE, given a longer span of time. The quality improvement in a *given time* is what differentiates a good CE environment from its precursors. However, CE should not try to snatch credit for the optimization attributable to the use some iterative design automation method. The quality improvement we are trying to measure is that which is made possible by different perspectives influencing each other's work in the CE environment.

## 14.6 CONCLUSIONS

Concurrent Engineering pursued is defined as a product development process conducted largely with the aid of computer networks on which certain common services are available to every member of the virtual team. Early detection and correction of design flaws and evaluation of more number of design alternatives are important benefits of CE. This helps in cutting down the product development time substantially due to the closer interaction of various working groups.

Because each organization may implement CE-based design uniquely there is no particular "how" of CE-based design. CE is at the heart of an organization's future competitive position. Also, the automated information support technology continues to change rapidly and very specific architectures will need to be evaluated over time.

In general, the implementation of CE should be forward planning based, i.e., it should be a constantly planned evolution towards a final vision.

## References

Landon C.G. Miller, *Concurrent Engineering Design - Integrating the Best Practices for Process Improvement*,

Concurrent Engineering - Technology and Standards, Tutorial #2., Tutorial (Briefing Slides) from *CE & CALS Washington '92 Conference*, Washington, DC, June 1, 1992.

Dwivedi, Suren N. and Sobolewski, Michael, Concurrent Engineering - An Introduction, *Proceedings of the Fifth International Conference on CAD/CAM Robotics and Factories of the Future '90*, Vol. 1: Concurrent Engineering, 3–16. New York: Springer-Verlag, 1991.

CERC Technical Report Series *CERC-TR-TM-92-002*.

Concurrent Engineering Transactions By-K.J. Cleetus and R. Reddy.

## Some Hyperlinks

- Concurrent Engineering Research Center
  http://www.cerc.wvu.edu/
- Society of Concurrent Engineering - Home Page
  http://www.soce.org/
- Concurrent Engineering and Shared Knowledge
  http://www.cad.strath.ac.uk/EIG/Concurrent.html

## CE-related Newsgroups

- Comp.groupware
- Comp.software-eng

# CHAPTER 15

# Product Data Management

## 15.1 INTRODUCTION

Manufacturing companies are usually good at systematically recording component and assembly drawings, but often do not keep comprehensive records of attributes such as size, weight, where used etc. As a result, engineers often have problems accessing the information they need. This leaves a wide gap in their ability to manage their product data effectively. *Product Data Management (PDM)* systems are used to manage both attribute and documentary product data, as well as relationships between them, through a relational database system. PDM can be defined as follows:

*"PDM is the integrated and computer supported management of heterogeneous information generated by different departments during every phase of product lifecycle: requirements capture, design, development, delivery and logistic support"* (Figure 15.1).

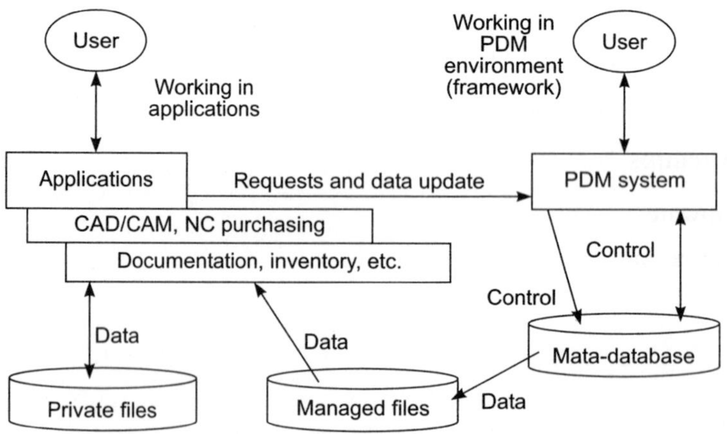


**Figure 15.1** Functional view of a PDM system

In order to achieve these goals, a PDM system will have the following two modules:

(i) Product data classification

(ii) Process management.

## 15.2 PRODUCT DATA CLASSIFICATION

Classification should be a fundamental capability of a PDM system. Information of similar types should be capable of being grouped together in different classes. More detailed classification would be possible by using attributes to describe the essential characteristics of each component in a given class.

### 15.2.1 Classification of Components

Components will be entered in the database under a variety of classes which suit the business needs. Classes themselves can be grouped together under convenient broad headings. This allows the company's working stock of components to be organized in an easily traceable hierarchical network structure. Each part can be given its own set of attributes. Additionally, some systems allow registering that certain components are available with specific optional attributes. This can be invaluable in controlling Bills of Materials (BOMs).

### 15.2.2 Classification of Documents

Documents relating to components and assemblies can be similarly classified; for example, classes might be 'drawings', '3D models', 'technical publications', 'spread sheet files', etc. Each document can have its set of attributes, viz., part name, part number, designer, date entered etc. At the same time relationships between documents and the components themselves can be maintained. So, for example, a dossier for a specific bearing assembly could be extracted, containing 2D drawings, solid models, and FEA files.

PDM systems vary greatly in their classification capability and they support the ability to define a classification only at the time when the database is implemented. In recent PDM systems these classes can be defined and modified at will as the demands of the organization change.

### 15.2.3 Product Structure

The third way product data can be accessed is by product structure. For any selected product, the relationship between its assemblies and the parts that make up these assemblies should be maintained. This would mean that one could open a complete BOM, including documents and parts, either for the entire product or selected assemblies. A distinct advantage of this method is the ability to hold not just the physical relationships between the parts in an assembly but also other kinds of structures; for instance, manufacturing, financial, maintenance or document relationships. This makes it possible for specialist team members to see the product structured from their point of view.

### 15.2.4 Querying the Data

The user should be able to get at the components and assembly data by a variety of routes. He could move up and down a classification tree; pick his way through a product structure; simply call-up the data he wants by searching for it by name or part number, or search for groups of data by specifying one or more attributes.

## 15.3 PROCESS MANAGEMENT

Process management is about controlling the way people create and modify data. While project management concerns itself only with the delegation of tasks, process management addresses the impact of tasks on data.

Process management systems normally have three broad functions:

(i) *Work Management:* This manages what happens to the data when someone works on it.

(ii) *Workflow Management:* This manages the flow of data among people.

(iii) *Work History Management:* This keeps track of all the events and movements that happen to the various functions during the execution of a project.

### 15.3.1 Work Management

The act of designing involves creating new data and changing data. A solid model, for example, may go through hundreds of design changes during the course of development, each involving far-reaching modifications to the underlying engineering data. Often the engineer will wish simply to explore a particular approach, later abandoning it in favour of a previous version design.

A PDM system offers a solution by acting as the engineer's working environment, meticulously capturing all new and changed data, maintaining a record of the version, recalling it on demand and effectively keeping track of the engineer's every move. When an engineer is asked to carry out a design modification, he or she will normally require more than just the original design and the *Engineering Change Order (ECO)*. Many documents, files and forms may need to be referred to by the members of the design team. In a traditional design environment one would compile a project folder or dossier, which the team could refer to as needed.

Current PDM systems cope with this requirement with varying degrees of success. One approach is to emulate paper-based processes by using what are known as 'user packets'. The packet allows the engineer to manage and modify several different master documents simultaneously as well as providing various supporting documents for reference. This approach also supports the concurrent engineering principle. For example, although only one user can be working on a master design, colleagues working on the same project can be instantly notified whenever the master design is updated, and reference copies of it will be made available to them in their own packets. A given packet can be worked on only by the user to whom it is logged out. But, its contents can be looked at and copied by everybody with the necessary access permissions.

### 15.3.2 Workflow Management

Packets have the advantage of making it easy for team members to share meaningful groups of documents, but they are useful for another reason too. They make it possible to move work around from department to department, or from individual to individual in logically organized bundles. During the development of a product many thousands of parts may need to be designed. For each part, files need to be created, modified, viewed, checked and approved by many different people, perhaps several times over.

Each part will call for different development techniques and different types of data - solid models for some, circuit diagrams for others and FEA data for others. Furthermore, work on any of these master files will have a potential impact on other related files. So there needs to be continuous cross checking, modification, resubmission and rechecking. With all these overlapping changes, it is all too easy for an engineer in one discipline to be investing considerable time and effort in pursuing a design which has already been invalidated by the work someone else has done in another part of the project. Bringing order to this highly complex workflow is what product data management systems do best. In particular, they keep track of the thousands of individual decisions that determine who does what next (Figure 15.2).

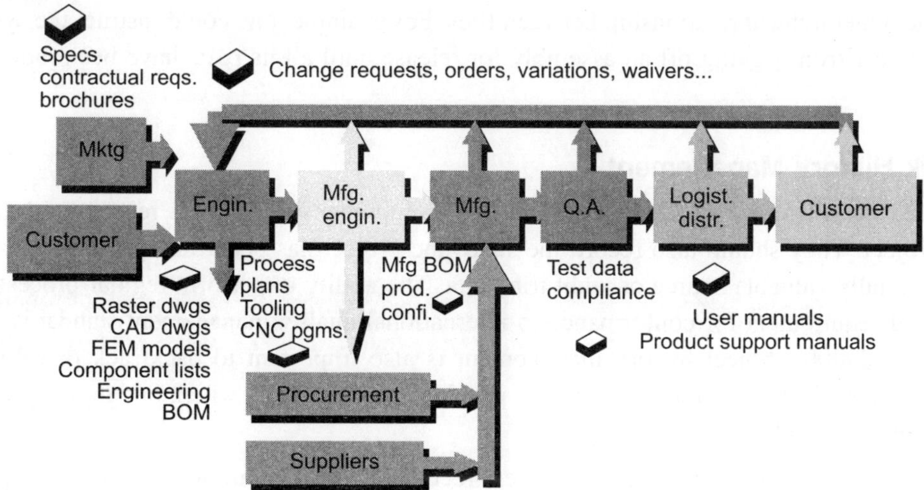


**Figure 15.2** Workflow in a PDM system

Most PDM systems allow the project leader to control the progress of the project via states using pre-determined triggers and a routing list, which may vary according to the type of organization or development project involved. The systems differ in the extent of flexibility they permit within the framework discipline. The most rigid systems are based on procedures. Every individual or group of individuals is made to represent a state in a procedure – 'Initiated', 'Submitted', 'Checked', 'Approved', 'Released'; a file or record cannot move from one individual or group to the next without changing states. Some systems make it possible to give the task an identity of its own, separate from the people working on it.

For example, suppose that an engineer working on a design wants to discuss with colleagues to find the best way to approach the design. So long as the master model and all the associated reference files are contained in and controlled by a packet, it is simple to pass the entire job across to any number of other people without triggering a change of state. The formal workflow procedure is uncompromised by this informal re-routing because the authority to change the file's state does not move around with the packet. It remains with the designated individual.

Communication within the development team is enhanced too. When packets of data and files are passed around, they can be accompanied by instructions, notes and comments. Some systems have alert capability; others even have provision for informally annotating files with the electronic equivalent of 'post-it' notes.

In other words, a process management system could be seen as a way of 'loosening up' the working environment, instead of constraining it. The challenge is how far one can allow informal teamwork and cross-fertilization to carry on and still keep overall management control of project costs and deadlines? Most systems allow the up-to-date status of the entire task, with all supporting data, to be tracked and viewed by authorized individuals at all times. A packet represents one task in a product development project, which may consist of many thousands. Each packet follows its own route through the system but the relationship between packets also needs to be controlled. To coordinate such a complex workflow effectively one needs to be able to define the interdependence of tasks so as to match the way individual project is structured. Not all systems are easy to customize in this way. The ones that are, have the ability to create a hierarchical relationship between files. For example, one could instruct the system to prevent an engineer from signing off an assembly for release until all its parts have been individually released.

### 15.3.3 Work History Management

As mentioned earlier, PDM systems should not just keep comprehensive database records of the current state of the project. They should also record the states the project has been through. This means that they are a potentially valuable source of audit trial data. The ability to perform regular process audits is a fundamental requirement for conformance to international quality management standards such as ISO9000 and EN29000. Project history management is also important to backtrack or rollback to specific points in a project's development where a problem arose, or from which one may wish to start a new line of development.

The specific development milestones the system records are important. Some systems record only the changes in ownership of documents. So ownership can be traced at a specific point in time, but not the modification itself. Others have the ability to record changes but may do so as a series of 'snap-shots' taken only when a file changes 'state'. This can leave large gaps in workflow history as a user may have been making modifications to a design for several weeks without any change to its state. Some systems provide an historic record, which is like a motion picture by allowing one to record changes to any system-defined level chosen - for example, every time a modified file is saved.

This level of historical tracking, as well as providing comprehensive auditing, also permits the active monitoring of individual performance which is invaluable in time-critical projects.

## 15.4 BENEFITS OF PDM

The benefits of PDM can be summarized in the following paragraphs.

**Reduction in Time-to-Market**

Reduction in time-to-market is the major benefit of a PDM system. It is achieved due to the following reasons:

- The three factors which control on the speed with which a product can be brought to market:
  (i) The time it takes to perform tasks, such as engineering design, and tooling

(ii) The time wasted between tasks, as when a released design sits in a production engineer's in-tray waiting its turn to be dealt with

(iii) Time lost in rework.

A PDM system can do much to reduce all these time limitations.

- It can speed up tasks by making data instantly available when it is needed.
- It supports concurrent task management.
- It allows authorized team members access to all relevant data, all the time, with the assurance that it is always the latest version.

## Improved Design Productivity

PDM systems, when driving the appropriate tools, can significantly increase the productivity of engineers. With a PDM system providing them with the correct tools to access this data efficiently, the design process itself can be dramatically shortened. Another factor is that designers should spend more time actually designing. Historically a design engineer would spend as much as 25–30% of his time simply handling information; looking for it, retrieving it, waiting for copies of drawings, archiving new data. PDM removes this dead time almost entirely. The designer no longer needs to know where to look for released designs or other data: it is all there on demand (Figure 15.3).

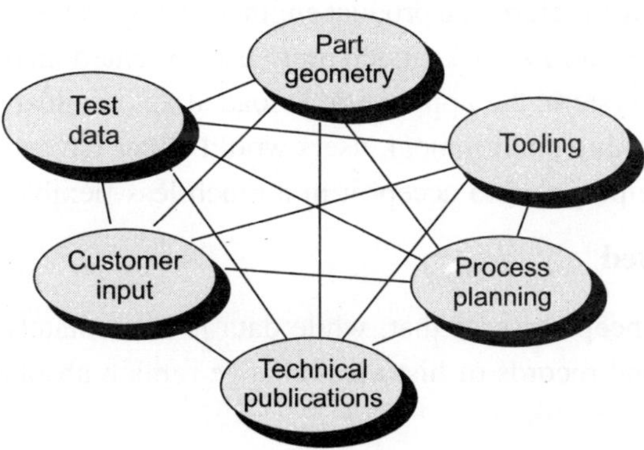


**Figure 15.3** Improving design productivity

A third major time saver is the elimination of the 'reinvent the wheel' syndrome. The amount of time designers spend solving problems that have probably been solved before, is well known. It is often considered quicker to do so it again than to track down design elements that could be re-used.

With a PDM system, however, the identification, re-use and modification of existing similar designs should become routine.

## Improved Design and Manufacturing Accuracy

An important benefit of PDM systems is that everyone involved in a project is operating on the same set of data, which is always up to date. If one is working on a master file he knows it is the only one; if he is viewing a reference copy, he knows it is a replica of the latest master.

So overlapping or inconsistent designs are eliminated - even when people are operating concurrently. Naturally this leads to far fewer instances of design problems that only emerge at manufacturing or QA, more right-the-first-time designs and, once again, a faster path to the market place.

### Better use of Creative Team Skills

Designers are often conservative in their approach to problem solving for no other reason than the high time penalties for exploring alternative solutions. The risks of spending excessive time on a radically new design approach, which may not work would be unacceptable. PDM opens up the creative process in three important ways:

1. It keeps track of all the documents and test results relating to a given product change, minimizing design rework and potential design mistakes
2. It reduces the risk of failure by sharing the risk with others and by making the data available to the right people fast
3. It encourages team problem solving by allowing individuals to bounce ideas off each other using the packet-transfer facility, knowing that all of them are looking at the same problem.

### PDM is Comfortable to Use

Although PDM systems vary widely in their levels of user-friendliness, most set out to operate within the existing organizational structure of a product engineering operation, without major disruption. The system should, in fact, make familiar tasks much more user-oriented than before. When users wish to view information on a PDM system, the application is loaded automatically, and then the document is loaded. In a conventional working environment, users would either have to be much more skilled at accessing the information or be prepared to accept it in a much less flexible form.

### Data Integrity is Safeguarded

The single central vault concept ensures that, while data is immediately accessible to those who need it, all master documents and records of historical change remain absolutely accurate and secure.

### Better Control of Projects

The reason that product development projects are almost invariably late is not that they are badly planned in the first place, but that they routinely go out of control. This is because the immense volume of data generated by the project rapidly snowballs beyond the scope of traditional project management techniques. The greater the competitive time pressures, the greater the scope for inconsistency, and likelihood of rework. PDM systems enable to retain control of the project by ensuring that the data on which it is based is firmly controlled. Product structure, change management, configuration control and traceability are key benefits. Control can also be enhanced by automatic data release and electronic sign-off procedures. As a result, it is impossible for a scheduled task to be ignored, buried or forgotten.

### Better Management of Engineering Change

A PDM system must allow to create and maintain multiple revisions and versions of any design in the database. This means that iterations on a design can be created without the worry that previous versions will be lost or accidentally erased. Every version and revision has to be 'signed' and 'dated', removing any ambiguity about current designs and providing a complete audit trail of changes.

### PDM is a Major Step Towards Total Quality Management

By introducing a coherent set of audited processes to the product development cycle, a PDM system should go a long way towards establishing an environment for ISO9000 compliance and Total Quality Management (TQM). Many of the fundamental principals of TQM, such as 'empowerment of the individual' to identify and solve problems are inherent in the PDM structure. The formal controls, checks, change management processes and defined responsibilities should also ensure that the PDM system contributes to conformance with international quality standards.

## 15.5 CONCLUSIONS

PDM integrates the various activities of product development as well as product lifecycle activities. This results in substantial reduction in time-to-market.

In principle, one can buy a PDM system from any vendor and use it to integrate the geometric, numeric and knowledge bases existing in the respective native formats of various vendors. However, in practice, this is not so yet. In other words, existing PDM packages are not generic enough. Therefore, almost every popular CAD/CAM/CAE developer also develops a PDM system which is fine-tuned with his other products. Table 15.1 shows a few developers and their respective CAD/CAM/CAE and PDM packages. For instance, if Ford use I-DEAS for CAD/CAM/CAE, they are constrained to buy Metaphase software package for their PDM. When developers adopt more open data formats, this difficulty can be overcome.

**Table 15.1** Popular CAD/CAM/CAE and PDM packages

| Developer | CAD/CAM/CAE | PDM |
|---|---|---|
| EDS | Unigraphics | I-MAN |
| SDRC | I-DEAS | Metaphase |
| PTC | Pro/ENGINEER | Windchill |

Another interesting fact is that there is a lot of commonality between PDM and ERP systems. While PDM is more inclined towards geometric database, ERP is more towards numeric databases. Therefore, it may be preferable to have a single package that provides PDM as well as ERP functions. Already the PDM vendors are trying to provide ERP functions and ERP packages are incorporating PDM functions. This trend will evolve into packages that provide ERP and PDM functions in an integrated manner.

Presently, considerable efforts are required for both ERP and PDM implementation in any organization as this calls for a lot of customization since each company has its own style of working and documentation. This is unfortunately not a one-time effort since the same may have to be repeated at least partly when these packages need to be upgraded. Often the cost of customization is more than that of the package itself depending on the size and style of the organization. Reengineering the company's working style and documentation before embracing to ERP or PDM might help in this regard.

❑❑❑

# CHAPTER 16

# Computer Integrated Manufacturing

## 16.1 INTRODUCTION

*Computer-Integrated Manufacturing (CIM)* involves the linking together of all business functions in a manufacturing organization to form an efficient unified enterprise. CIM is not confined to the shop floor, but embraces the total business. CIM systems provide the communication links to bring together the numerous computer tools into a cohesive network, which in turn provides prompt up-to-date information for decision-makers and the control functions to implement their decisions.

CIM should not be implemented by installing customized links within the organization; it rather calls for reengineering of the organization. The process should start with a basic appraisal of the total business before moving on to looking at the long-term plan for the provision of facilities such as equipment and network selection. The planning and implementation of a CIM project is through a top down and bottom up process. An approach that is taken from software engineering is like more of a waterfall model.

The components required for CIM are available and in use; however, there is no one supplier who provides the complete CIM system. This leaves users with the problem of combining hardware and software from heterogeneous suppliers into an integrated system. Many organizations, therefore, spend a lot of time and money in investing in expensive interfaces and customization. Over the years standards have developed and integrated within manufacturing sectors.

According to James Clark's, Author of *CIM Planning*, CIM promises a world in which manufacturing has been transformed through the use of technology into a truly competitive weapon. Indicative of this are the following benefits directly attributed to CIM:

- 5 to 20 percent reduction in personnel costs.
- 15 to 30 percent reduction in engineering design costs.
- 30 to 60 percent reduction in overall lead times.
- 30 to 60 percent reduction in work-in-process.
- 40 to 70 percent gain in overall production.

- 200 to 300 percent gain in capital equipment utilization.
- 200 to 500 percent gain in product quality.
- 300 to 3500 percent gain in engineering productivity.

### 16.1.1 Definitions of CIM

CIM is a tool to be used to achieve or facilitate defined corporate goals or strategies. It is a way of doing business that emphasizes an automated coordination of information and effort throughout all the functional areas of a corporation. CIM has been uniquely and authoritatively defined by every major computer manufacturer and consulting firm in the past few years, and their definitions understandably have emphasized particular strengths and capabilities of the systems or services they provide. CIM also can be considered as a new important trend in the field of using information technology to improve manufacturing industry. CIM require the total Management system in which it can combine the development of product and its related business activities through the utilization of networks. The objective of CIM can also be aimed at lead-time reduction, productivity improvement, and "quality and reliability" control for future cost reduction. CIM is becoming very popular in automated assembly lines mainly because it can introduce dynamic management through the direct connection of production and sales.

The definition of CIM immediately points out three areas for consideration:

(i) CIM requires management to clearly express corporate goals and manufacturing strategies. If these goals or strategies are not established and expressed, there can be no rational basis for committing resources to CIM or potentially, resources could be allocated to initiating CIM in areas of no long-term interest to the corporation. Since CIM is a way of doing business, there must be a total commitment to CIM by management at all levels.

(ii) CIM is expensive in terms of hardware costs, software development, and requires dedicated personnel for the planning and implementation phases. A typical corporate CIM plan may take as long as five to ten years to fully implement.

(iii) CIM facilitates both access to and exchange of information. In many cases, CIM cuts across historical and traditional departmental lines of responsibility and may be viewed as invasive.

Figure 16.1 depicts various market conditions. These keep changing. Because of these frequent changes in a buyers market, the life of products keeps coming down as illustrated in Figure 16.2. It can be seen that the product is almost obsolete even as it enters the market. Therefore, shortening of time-to-market is vital for the survival of organizations. In other words, organizations have to react quickly to changes in market demands, more so in the global economy. More and more manufacturing organizations are integrating computer technology into their operations to achieve this. CIM is concerned with providing computer assistance, control and integration to all levels of the manufacturing industries, by linking automation into a distributed process system. The technology that is applied in the use of CIM makes intensive use of distributed network of computers, data processing and an effective database management system.

**Figure 16.1** Present market conditions

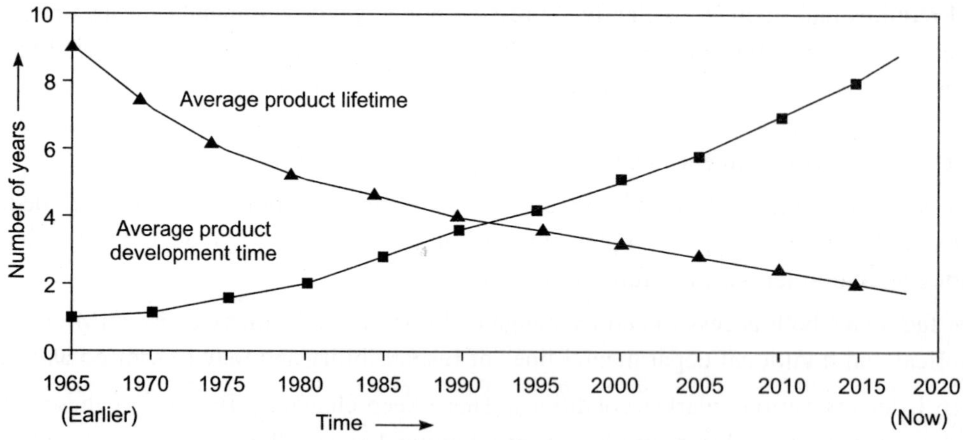


**Figure 16.2** Product development time vs. Product life time

CIM concerns itself with information retrieval at all levels of the production factory and its management. The term CIM has widely become accepted for linking together different applications through the use of computers in manufacturing in such a way that information can be shared across and transferred among these applications. CIM can be heralded as a "paperless factory" of the future.

CIM is a unified network of computer systems performing and/ or controlling the total integration of a business function. CIM integrates automated systems into the processes of producing a consistent products; it can be seen as the logical organization of individual engineering processes, production and market support functions into a computer-integrated system.

CIM is a way of bring together computer power to bear on the four main functions of manufacturing:

(i) Product design

(ii) Process control

(iii) Planning

(iv) Control systems.

The concept of CIM encompasses an integrated database that flows through the entire process from design to financial control, with all transactions - such as production, process, automatic process controls, storage, maintenance - all being driven from the same database.

CIM is the integration of CAD, CAM and production management systems.

These are just basic definitions about CIM. By reading all the definitions, it is clear that CIM encompasses all the processes of manufacturing. It allows for the retrieval of information at any stage of development, be that planning, finance or end product. The use of CAD and CAM allows CIM paradigms to control the overall architecture of a product.

CIM covers all aspects related to manufacturing business, including:

- Evaluation and development of different processes.
- Market analysis and future forecasts.
- Generation of future systems through analysis of product and market characteristics.
- Design and analysis for the processes relating to the manufacture of the required components (i.e., welding, cutting, painting etc.).
- Determination of the capacity of production relating to a particular project.
- Economics analysis of the total system and its disturbances.

The list above indicates that CIM aims to let information processing technology to penetrate into all areas of manufacture in order to:

- Make the entire manufacturing process more productive and efficient.
- Increase product reliability.
- Reduce maintenance cost related to manufacturing system and the product.

As more and more manufacturing businesses begin to integrate computers into their operations, then a paperless innovation culture will emerge. The new culture will allow information to be retrieved, sent and viewed for a product at any given stage of its life cycle. The use of CIM systems will provide a heavy reliance on organizations in using an active and efficient database system that allows the free flow of information quickly and with ease.

CIM is very efficiently revitalizing manufacturing processes and their industry. As CIM concentrates on data information flow and retrieval, its overall business philosophy is geared towards integrated manufacture. This forces out any unnecessary waste by reducing complexity and cost whilst improving quality to become more competitive.

There are three prerequisites for CIM implementation in an organization:

(i) All activities must happen on computers.

(ii) There must be reliable connectivity among these computers through LAN/ Internet/ intranet.

(iii) There must be well established communication protocols among the computer systems. Furthermore, there shall also be a reliable security system to monitor and control the data that may be distributed across the network.

## 16.2 COMPONENTS OF CIM

The following are the components of a CIM System:

(i) *Computer-Aided Design (CAD):* CAD is a core module in the CIM system and has been widespread for many years. CAD means computer assistance while the human operator converts his/ her ideas into a graphical or a mathematical model. The key to CAD is its conversion from a model represented in the human brain into the machines in the form of digital data.

(ii) *Computer Aided Manufacture (CAM):* CAM is seen as a system, which supports production engineering by defining operating sequences, establishing requirements for fixtures, tooling and simulating operations prior to "first-off" machining. CAM takes a focus towards *Flexible Manufacturing Systems (FMS)*. These consist of two or more machine tools linked by an integrated part handling system and operating under the control of a computer. The control system can route parts between machines so as to achieve maximum utilization of the manufacture process.

(iii) *Automated Materials Handling:* In order for materials to move efficiently during manufacture, full integration must be allowed. Automated handling systems include cranes, Automated Guided Vehicles (AGVs) and robots.

(iv) *Group technology (GT):* This is classification and coding of parts in order to ensure that manufacture of similar products is done in the same way. The benefits of GT:

- Design retrieval: The rapid identification and retrieval of information.
- Cellular Manufacture (CM): The creation of cells to manufacture identical parts.
- Purchase support: Association of related parts.
- Service depot streamlining: Identification of alternate parts and reduction of variety.

(v) *Manufacture Resource Planning (MRP):* This deals with the inventory control, ordering, bills of materials, but does not address the overall requirements of the business.

The elements listed above are the main building blocks of CIM. The main aim of a CIM is to enable a manufacture organization to retrieve information about its manufacture process efficiently with speed and accuracy. In any organization information is retrieved through an effective database system that operates through a network.

## 16.3 BARRIERS TO CIM IMPLEMENTATION

There are several major obstacles to the acceptance of CIM as a practice management solution. The following are five of the most frequently encountered:

(i) *CIM is an expensive undertaking that can only be done by those with the finance to afford it.* Most managers see CIM in terms of its technology. This technology is inevitably expensive, not easy to implement and requires huge resources for testing and debugging the systems. In addition, these investments are seen as ones that must be made upfront. For many firm these costs represent a significant barrier.

(ii) *The Implementation of CIM is a long-term undertaking*. When most managers look at the experiences of firms that have been successful with CIM, they see endeavours that took many years to complete. For a company that is faced now by dropping sales, falling margins, and rapidly eroding profits, there is little attraction in examining a system that may require four or five years before it yields any benefits.

(iii) *There are cheaper alternatives to CIM available*. For many managers who are interested in improving manufacturing performance and enhancing corporate competitiveness, there are other alternatives available such as system simplification, time-based competition, competing on quality, and just-in-time manufacturing. What makes these alternatives so appealing is that they appear to require less investment.

(iv) *There are relatively few success stories*. The concept of CIM has been around since the early 1970s. During this period, many firms across the country have attempted to take advantage of the potential benefits of CIM. Firms have tried to implement technology such as FMSs, robotics, MRP, NC machines, AGVS, and CAD systems. However, very few of these attempts have succeeded. The success stories in CIM are both few and less publicized.

(v) *The fear of loss of management control*. For many managers, CIM represents a potential loss of management control. These managers tend to understand the business of building and delivering their firm's products. They tend to have "hands-on" experience. They have been with their companies for a long time. Often, their authority in their firms is based on their knowledge of the firm. CIM represents a major challenge to their position and authority within the firm. By introducing automation and technology, CIM often redefines the production process. It changes how the firm plans, schedules, executes, and tracks production. Much of the skills and knowledge possessed by the current management may no longer be pertinent to this new CIM-run environment. These managers may feel that they no longer control the operations in this climate.

## 16.4 A CASE STUDY IN CIM

The Container Division of Continental Can Corporation is located in West Chicago, Illinois, U.S.A. They manufacture can-making and can-closing equipment and parts. Their major customers are bottling and beverage companies such as fruit and soft drinks. Their customers are from local small factory to large regional distributors and bottlers that supply the well-known fruit and soft drinks. Its major competition comes from two companies American Can and National Can.

In 1983 the Container Division was in serious difficulties. It was losing both money and market share. Finally, it had developed a reputation for poor service. For example, to get a spare part from Continental Can would require a wait of months, not weeks. Some of its customers had become so frustrated by the long waits that they had started turning to its competitors for spare parts.

Based on those situations Continental Can, had decided that something had to be done about the situation at the Container Division. After 12 weeks research a consultant company submitted a report. The consultant company concluded that the problems that were currently plaguing the division could be easily identified and costs could be reduced. The factory could be improved by adopting CIM.

### Move to CIM

After assessing the problems, it was decided that CIM offered the most advantages for the Container Division. A proposal was submitted to the executive board of the Continental Can Corporation. The primary justification for the proposal was the anticipated cost savings resulting form improved throughput and reduced inventory.

In approving the proposal, top management recognized that this project was essentially a long-term project. They agreed to commit the required investments. Specifically, top management indicated that they were willing to allocate the money needed for the acquisition of any necessary technology. They also agreed to use a long-term perspective when evaluating the division's performance. Finally, they agreed to protect the project and the implementation team from outside pressures.

### Implementing CIM

Within three months the project was approved and ready for implementation.

Because of the limited time availability of certain key management personnel, it was decided by everyone involved in the implementation that the Container Division had only one chance to get it right. To improve the chances of success, top management insisted that certain guidelines be followed in the implementation:

(i) *Training and education:* Everyone in the division, from top management to the labour, was to be exposed to the CIM system. So, Training and education were very important and emphasized in this project.

(ii) *Resistance to change was not to be tolerated:* Education and training was considered important and required for the success of CIM.

(iii) *The CIM implementation was to be focused:* It was decided that the changes demanded by CIM were to be confined to the manufacturing system and the information system supporting its operation.

(iv) *Manufacturing basics were to be improved:* Attention was devoted to the areas of master production scheduling, production planning, materials planning, and shop floor control. These were areas that were currently being done very poorly.

(v) *The manufacturing system was to be protected as much as possible from external changes:* External changes were viewed as very disruptive. The system was to be protected in three ways. The first was through the master production schedule. Time limits were introduced. An order review and release mechanism was introduced to manage the flow of work from the planning system to the shop floor. Finally, extra capacity was added on the shop floor. This final policy reduced the potential impact of any bottlenecks. It also simplified the entire process of capacity planning.

(vi) *System simplification was to precede system automation:* Because of difficulties with work flow and shop layout, management decided that these problems were to be eliminated by changing the layout of the floor and straightening the flows.

(vii) *Manufacturing facilities were to be consolidated:* There was no real need for different facilities. All production was to be concentrated at one plant.

(viii) *Use generalized systems wherever possible:* People had to change their practices to conform to the requirements of the new systems to reduce the implementation lead-time. Given the limited availability of certain critical management personnel, any activity that lengthened the lead-time was to be avoided.

(ix) *Get the right people involved and committed:* The new CIM system had to be a system that everyone believed in and used. There had to be ownership by the users of this new system. To achieve this goal, systems people were treated as support. They were not to lead the implementation effort. That task was to be left to the users.

These guidelines provided the general structure within which the implementation took place. From the perspective offered by hindsight, they also significantly simplified the implementation process by identifying in advance many of the potential pitfalls in the implementation process.

**Tangible and Intangible results**

CIM system is showing signs of success. The effectiveness of this new system could be found in the following benefits now accruing to the division:

- Lead times had fallen by 40 percent.
- Inventories had fallen by some 33 percent, with a 50 percent reduction being expected.
- Earlier only 25 percent of all orders were delivered on time. Now this had risen to 91 percent, with 95 percent expected soon.
- Data accuracy had increased significantly. Bills of Materials are now 96 percent accurate. Routings are 97 percent accurate. The accuracy on cycle counts averaged about 95 percent. Furthermore, errors, whenever encountered can be corrected quickly.
- The lead time on spare parts dropped from nine months to about one month.

The division is making profit. It is now considered to be competitive in the marketplace. Its reputation for service has greatly improved. The company has become so successful in the spare-parts segment of its business that management is beginning to consider supplying spare parts for equipment produced by its competitors. Now the company could be regarded as being the major competitive force in the market.

In addition to these tangible benefits, there were other equally important intangible (indirect) benefits:

- People involvement had increased significantly. People were now making decisions and taking responsibility for their actions.
- Cross-functional communication and coordination was increasing.
- The communications with suppliers had improved. The Container Division was now able to give its suppliers better schedules. As a result, the division was now able to get better and firmer commitments from its suppliers to the schedules. The Container Division was also able to reduce the number of suppliers to some 500.
- Confidence in the CIM system and in the overall ability of the Container Division had improved significantly. The planning was better. People were now able to get good product cost information from the CIM system in hours or days, whereas earlier it would take weeks or months. Because of this ability to get cost data quickly, the Container Division was able to compete on its ability to quote rapidly.
- Better data was leading to better actions.
- The overall operation environment had improved greatly. The level of stress previously experienced was now disappearing. People no longer blamed each other for the problems taking place everywhere in the system.

As a result, CIM had succeeded at the Container Division. Most of the goals initially set for the CIM project had been achieved. Because it was able to effectively support business activities in this division, everyone agreed that CIM was a strategic project.

## 16.5 DEVELOPMENTS IN CIM RESEARCH

Today, we are at the threshold of seeing rapid changes in information processing and communication equipment. While the mass production is still needed for certain sectors of manufacturing, it is no longer a competitive weapon in a rapidly changing global economy. The challenge in reviving the manufacturing industry to face the competitive threat depends on the ability to respond to all the requirements of the global market. However, many of the manufacturing industries in the developed world are still operating islands of automation or old age technologies. This trend of clinging to the past will not allow the manufacturing industries to face the growing competition in the global market. Today many companies are relocating offshore to reduce labour costs. This phenomenon can be avoided if manufacturing companies invest in CIM and related technologies and become globally competitive. The efficient application of CIM in a fully integrated manner will help the industries face the global and local competition with a high degree of confidence. With the advent of new network technology and Internet, the application of CIM has transcended geographical boundaries and embraced the global application of CIM. To facilitate the investment in CIM and to help manufacturing industries, many researchers are seeking solutions in application of CIM. Some of the research trends in CIM which are evolving today can be categorized as follows:

- Justification of CIM and management strategies for CIM.
- Enterprise integration for CIM beyond and within geographical boundaries.
- Network communications for the implementation of CIM.
- Advanced tools and technologies for the application of CIM.
- Manufacturing system modeling.
- Application of Artificial Intelligence (AI) such as fuzzy logic, neural network, genetic algorithms and intelligent agents for fully integrated *Intelligent Manufacturing Systems (IMS).*

Each of these categories is diverse in nature and consists of many sub groups. In addition, a few research directions may overlap in more than one group due to the complexity of the research in CIM.

Justification and management strategies for CIM consist of multitude of research directions. This research trend focuses on providing managers with principles and guidelines for the application of CIM in their working environment and approaches to overcome the resistance to change towards integrated manufacturing. Further, this categorization addresses the organizational and social issues in an enterprise, issues involving benchmarking the features of the manufacturing supply chain and manufacturing strategy implementation through real time systems, optimized enterprise integration and decision support systems (DSS) and Expert Systems (ES) for manufacturing managers. DSS and ES may further include ES for quality control, ES for production planning and selecting manufacturing parameters, DSS to evaluate designing decisions, financial decisions, and educational and informative systems.

Enterprise integration for CIM beyond and within geographical boundaries consists of a research diversity, which includes architectures and modelling for enterprise integration, evaluation methodologies for enterprise integration, and international CAD/CAM collaboration for CE and CIM implementation through integration of subsystems.

Network communications for the implementation of CIM category include application of wide-area networks and Internet for CIM, information enhancement by data integration, issues related to integration of client and server for manufacturing shop floor automation, application of multimedia and hypermedia in CIM environment, and data management in CIM systems.

The categorization of advanced tools and technologies for the application of CIM is one of the thriving areas in CIM-related research. It consists of robotics and automation for a CIM framework, vision-based manufacturing systems and intelligent AGVs and intelligent and knowledge-based enabling systems for CIM.

Manufacturing system-modeling category consists of research variations, which include integration of information models with functional models for CIM, object-oriented resource modeling, integrated simulation modeling approaches for CIM, and modeling approaches and methodologies for designing CIM systems.

Application of AI for fully integrated IMS consists of research directions, which include the application of neural network on manufacturing automation and technologies, intelligent scheduling systems with genetic algorithms, fuzzy scheduling systems, hybrid systems of CIM components, adaptive models for CIM architectures and embedding AI in CIM software.

## 16.6 FUTURE DIRECTION OF CIM

In today's competitive global market survival of any industry depends on its ability to communicate and transfer the right information at the right time to the right people. Manufacturing cannot escape from this present requirement. Having an ability to communicate for effective management and manufacturing activities across the geographical boundaries among the globally distributed resources will significantly benefit manufacturing industries. Today a number of global conglomerates are formed in many facets of industry. A virtual enterprise is a network of interconnected global conglomerates. Predicting the future research direction of CIM and related areas is a difficult task in ever expanding and growing technological development era. However, an attempt is made to foresee the future direction, which will dominate the researchers' minds for the next decade, based on the current developments in CIM research.

Today's competitive and agility requirements of the global market can be only met by *virtual enterprises*. To provide a better future in the present market requirements, research in *virtual CIM* and the application of it in worldwide manufacturing industries are beginning to emerge (Figure 16.3). Application of virtual CIM has been proposed as a necessary step towards the future in manufacturing to face competitive challenges. However, many development works need to be carried out to face challenges faced by virtual enterprises. Hence, the research should be further strengthened towards developing a virtual CIM to satisfy the globalized and distributed manufacturing enterprises of today in order to meet the competitive and agility requirements of present market conditions.

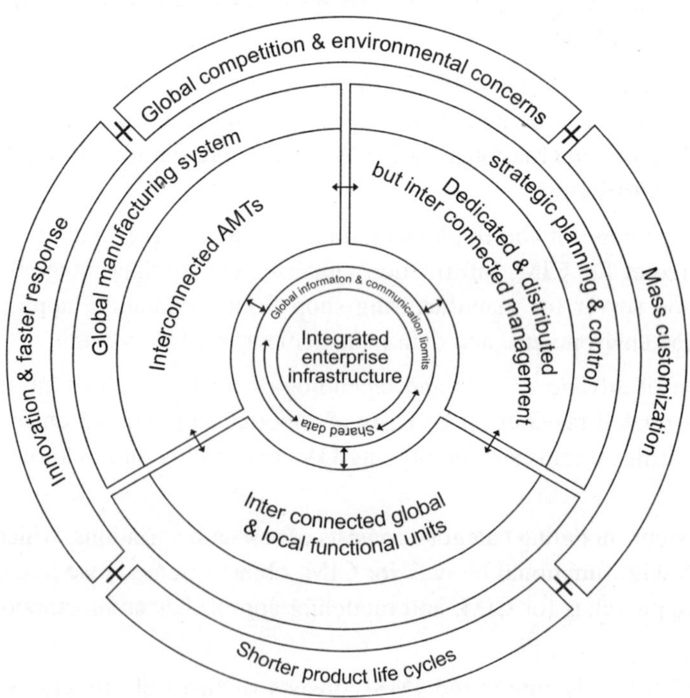


**Figure 16.3** Virtual CIM wheel representing the global market conditions

In a virtual enterprise the integration of information is praised, as only through information can a virtual organization become meaningful, and only by selecting a new generation of information technology can this vision be realized. To represent the evolving process of CIM and to reflect the present need for a virtual CIM, which stresses the importance in strategic and integrated management in implementing CIM in globally distributed enterprises.

In the justification of CIM and management strategies for CIM research area, recent developments in the group-based management information systems have strengthened the application of a DSS in a group decision environment. In addition, rapidly changing technology and the increasing reliance on knowledge and communication create significant challenges to static and old economic models of an enterprise. Therefore, in order to justify and optimize investments in a virtual enterprise, a need exists for a *Group Decision Support System (GDSS)*, which has a capacity to consider the selection of appropriate *Advanced Manufacturing Technologies (AMTs)* in a global framework. A GDSS can improve decision outcomes by using structured approaches for the unstructured problems in a group decision-making endeavour. A GDSS will help decision-makers in a team environment, analyze the investment decisions of a company in an interactive manner, either in geographically distributed or a localized manufacturing enterprise with diverse functional units.

The rapid developments in technology provide innovative approaches to manufacturing and the recent trend towards global manufacturing has increased the number of geographically distributed and multi-location enterprises. In the research direction of enterprise integration for CIM beyond and within geographical boundaries, a collaborative framework for remote machining and planning and control systems have been proposed. The prime technology, which propels the advances in manufacturing system and international networking, is the communication technology. Hence, the research direction of network communications for the implementation of CIM has produced a number of research articles. In addition, the use of Internet and the World Wide Web in manufacturing applications is gaining momentum. However, research should be further strengthened towards providing integrated architectures and communication protocols for virtual CIM framework to enable real-time machining, processing, and decision-making facilities in geographically dispersed enterprises.

Research in robotics and AGVs is predominant in the research direction of advanced tools and technologies for the application of CIM. In today's technologies the remote manipulation of these machines has already been achieved. Mobile robots and global positioning satellites have played a significant role in remote manipulation of robots. Collaborative robots are being developed to machine complex surfaces and large components. In addition, vision-based systems have been developed towards Rapid Prototyping, part identification and processing. These developments need to be enhanced towards providing practical solutions for a globally dispersed manufacturing enterprise and be tested in a virtual CIM framework.

Today, many manufacturing industries are under pressure to achieve their objectives within limited capital available for new investments. Optimization of resource allocation for strategic investments could provide a mechanism to achieve the goal of the enterprise. The future of CIM depends on optimized integrated systems. Traditionally, to improve the effectiveness and competitiveness of an organization, all the efforts are focused on achieving a single goal or achieving the objective within a functional unit. However, the traditional method of single goal seeking mechanism or optimizing the results of an individual unit does not satisfy the overall objective of a manufacturing company. Therefore, a multiple goal seeking or multicriteria optimization which considers all the relevant factors and overall benefits of a company as a single unit, has been proposed to make CIM an important application in the present global economic circumstances. The existing multicriteria optimization mechanisms should be further enhanced or new approaches be developed for application in the globally distributed virtual CIM enterprises, so that the benefits of virtual CIM is fully realized. In addition, global integration approaches, which regard a globally distributed company as a single entity, should be developed and utilized to make virtual CIM an important application to meet the present global economic circumstances. Intelligence manufacturing is the path for the future. Therefore, manufacturing technologies should be blended with intelligence, which will help manufacturing enterprises to produce better quality products and manufacturing equipment to solve problems encountered during the normal course of operation itself.

❑❑❑

# Conclusions

The authors have made an attempt to give an overview of various ways of developing a product at the least possible time. They call it as *Rapid Product Development (RPD)*. In fact, this is not a new term; one can find a number of web sites with this title although they address only a segment of RPD. Many universities, technical institutes, manufacturing industries and R & D centres now have RPD Laboratories.

RPD is a synergic integration of virtual and physical technologies. Synergy of the activities involved is essential to carry forward the benefits achieved at any one activity till the end. For instance, when a human operator with a variable cycle time and a robot with a constant cycle time are in series production, the variability of the production line is much more than that of the human operator as the robot may be idling at times waiting for the operator to finish. The situation can be corrected by replacing the operator by an automated system or by introducing a buffer in between them.

In today's economy driven by global competition, RPD makes more sense than ever before. Industries no longer can afford to remain complacent with their current performance. They have to continuously improve to stay in business. For this, adopting new technologies is essential. Often, they invest in new technologies such as CAD/CAM but do not train the employees adequately. There is a misconception that value addition to the employees is not value addition to the organization as he may leave any time. With the less skilled engineers, the new technologies cannot not be exploited to the fullest extent. Continuous upgradation of the knowledge and skill of the employees, and keeping them satisfied with their working conditions and social life are very essential for the success of the organizations.

❑❑❑

# Index

## E



## F



## G



## H



## I



## L



## M

### T



### U



### V



### W